THE RCM Solution

THE RCM Solution

A Practical Guide to
Starting and Maintaining
a Successful RCM Program

Industrial Press, Inc.

Library of Congress Cataloging-in-Publication Data

Regan, Nancy.
The RCM solution : reliability-centered maintenance / Nancy Regan.
p. cm.
Includes bibliographical references and index.
ISBN 978-0-8311-3424-2 (hard cover)
1. Maintainability (Engineering) 2. Reliability (Engineering) I. Title.
TS174.R443 2011
620'.0045--dc22

2011004061

Industrial Press, Inc.
32 Haviland Street, Suite 3, South Norwalk, CT 06854
Tel: 203-956-5593, Toll-Free: 888-528-7852
Email: info@industrialpress.com

Sponsoring Editor: John Carleo
Copyeditor: Robert Weinstein
Interior Text and Cover Design: Janet Romano

books.industrialpress.com
ebooks.industrialpress.com

To my mother

For all the walks on the beach that brought me closer to this place, especially the one twenty years ago that started it all.

To the late John Moubray

For teaching me to Tango when everyone else was doing the Waltz

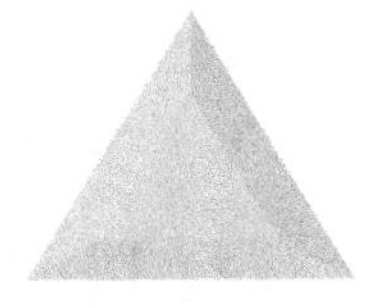

Table of Contents

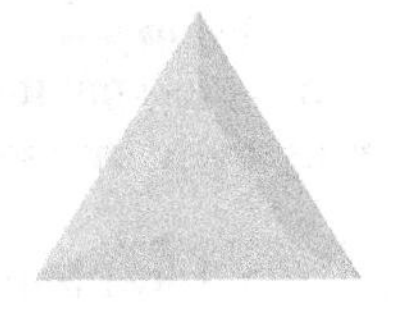

Acknowledgements

There are so many people who have come into my life offering friendship, guidance, advice, and support that have made this book possible. Your significant impact is woven into these pages. Some of you I am no longer in contact with, but the spirit of what you brought to my life and to my work remains with me always.

To my beloved husband, Dr. Dennis Fernandez, for embracing my dreams as you do your own.

To my Mother, whose love follows me wherever I go, whatever I do. Your job is not yet done.

To Sister, my sibling, my confidant, and my friend. Thank you for your unfailing encouragement, love, and understanding. I love you.

To my memory and my best friend in all the world, Mitchell Friedman. Your friendship, encouragement, and collaboration are some of the greatest blessings of my life. What would I do without you?

To the late John Moubray for being instrumental not only in learning my craft, but for your encouragement and guidance in launching The Force.

To Christina Thomas-Fraser for lighting the way.

To Michelle Duvall who proves every day that not all angels have wings.

To Campbell Fraser for your guidance. I am smarter, wiser, and more aware because of you.

Acknowledgements

To Mark Samuelson. From the first moment at Fort Campbell your kindness, patience, friendship, and advice mean more to me than you may ever know. Thank you for every mile we ran together.

To John Sims for sticking with me through long facilitation sessions and for being such a staunch advocate of our work.

To the CH-47 Cargo Helicopters Project Management Office, Redstone Arsenal, Huntsville, Alabama. Thank you for your unfailing support of RCM and The Force, and for your willingness to use RCM in so many unique and varied ways.

To Julie Haralson for your dedication to The Force, Inc. and your assistance with this manuscript.

To Colonel Jay Montgomery, USMC, for recognizing what RCM can do for asset management and for promulgating RCM wherever possible.

To AFCM Sam Campbell (Retired) for teaching me to never forget where I came from. The experiences you provided shaped my understanding of how critical the skill and knowledge of the user and maintainer are. I carry our friendship in my heart.

To Ralph Holland for teaching me so much and for making me laugh until I cried every step of the way.

To Steve Little for inspiring me to think beyond the normal bounds of RCM.

To Dr. Mark Horton for your generosity, consistent support, and precision regarding Failure Finding. The insights and counsel you offered me go far beyond mathematics.

To Malcolm Regler for giving me the "skinny" on all things RCM.

To Andrew Matters for helping to expand my knowledge and understanding of RCM.

To the Naval Air Systems Command (NAVAIR) PMA-260, Common Support Equipment, Lakehurst, New Jersey, for giving me my start. Special thanks go to Marie Greening and Maris Gultnieks.

Acknowledgements

To Rear Admiral Michael D. Hardee (Retired) for your support and encouragement and for teaching me to remember who we really work for.

To Tom Bleazey, NAVAIR Lakehurst, for supporting the first pilot projects...and beyond for NAVAIR PMA-260, Common Support Equipment.

To Diane Pullen and Cathy Malvasio. I hope you know how much your friendship and support meant.

To John Carleo, Editorial Director, Industrial Press, Inc. for your guidance, gentle understanding, and encouragement. It has been a sincere pleasure.

To my editor, Robert Weinstein, Gerson Publishing Company, for your honest, clear, and to-the-point edits. You made my final version of the manuscript easy!

To Janet Romano for diligently working on graphics and layout.

To Brandon Lee for staying by my side and keeping me on track.

To Charlotte for clearing my head when I really needed it.

To the nesting barn swallows that kept me company outside my office window as I finished this book.

Most of all, I'm grateful to the divine presence that leads and inspires this work.

Preface

Reliability Centered Maintenance (RCM) is a time-honored, proven process that has been employed all over the world for over four decades in nearly every industry. Because its principles are so robust, so powerful, and so versatile, the process has stood the test of time and human meddling. There are many RCM processes on the market that embody different approaches; many of them depart significantly from what was intended by the original architects of the process, Stanley Nowlan and Howard Heap. The basic principles of RCM have been criticized and manipulated because it is often wrongly believed that RCM takes too long to perform, or it's too expensive, or all of the steps are simply unnecessary. This just isn't so. RCM is a majestic process that gives an organization the opportunity to transform into a more safe and cost effective institution. However, the process must be performed *correctly* by the *right people*.

This book is intended to be a straightforward, no-nonsense presentation of what RCM is and how it can be applied to maximize results. *The RCM Solution* embodies minimal theory. Instead, it embraces the majesty of RCM's basic principles and sets forth a very common sense approach to achieving powerful results. As retired British Royal Navy (RN) Commander and former head of the RN's RCM program Andrew Matters once told me, RCM is nothing more than "common sense applied to physical assets." In that spirit, this book espouses exactly that. *The RCM Solution* is intended to be an introduction to RCM principles.

Anyone who knows me knows that I am an RCM zealot. I sincerely believe that if an organization chooses to employ RCM, it should be done so correctly. Nevertheless, as I gained more experience in the field, I came to understand that RCM cannot be done on all equipment simply because there aren't enough resources to do so. I also realized that not all assets require the rigor that RCM embodies. It was then that I embarked upon formulating less robust tools that can play a significant role in an organization's transformation. Therefore, this book introduces other asset management processes that embody RCM principles. But they are just that—*other processes*. They are not RCM and should be used responsibly—not as an excuse to use an alternative to full-blown RCM.

The intention of this book is to cut through all the noise, marketing, and false information about the process and simply set forth the principles of RCM in a way that can inspire organizations in starting and maintaining a successful RCM program. I have facilitated RCM using the techniques introduced in this book on assets ranging from plant equipment, mobile ground equipment, to aircraft for over 13 years. What I *know* is that if the principles of RCM are used *correctly* with the *right people*, the results can be transformative. I have seen it first hand.

During my RCM practitioner training I asked a question of my mentor, the late John Moubray, when he was presenting RCM theory because I thought I detected an inconsistency in the lesson. He simply looked at me and said "listen to the music, not the words" and he quickly moved on. The lesson was to recognize the versatility of RCM. RCM principles are like the paints that an artist uses. The same paints used by ten different artists will produce ten different paintings. How the paints are used is what determines if a masterpiece unfolds. So too are the principles of RCM.

This book comes from my heart. Its words are not only what I believe to be true, but what I know to be true. John trained his network members to be responsible custodians. He said it best when he affirmed "we are here to promulgate the principles we believe to be best practice and in so doing make the world a safer place for all who live in it."

During another conversation, John described how RCM principles could be used better in a particular industry. When I asked him why they weren't doing it better, he responded by simply saying "because they're doing the Waltz." When I asked him why they were doing the Waltz, he said "because no one is playing the Tango."

May I have this dance?

Nancy Regan
Madison, Alabama
May 2012
nancyregan@theforceinc.com

Introduction to Reliability Centered Maintenance

I am always excited to discuss Reliability Centered Maintenance (RCM) because I have seen first hand the overwhelming positive results that can be reaped when the process is applied correctly with the right people. RCM isn't a new process. The application of its principles spans four decades; it has been (and is being) applied in nearly every industry throughout the world.

Contrary to criticism about the process, RCM can be carried out swiftly and efficiently when executed properly.

RCM principles can be widely applied to an entire asset or more narrowly applied to select pieces of equipment.

RCM is one of the most powerful asset management processes that can be employed. Contrary to criticism about the process, RCM can be carried out swiftly and efficiently when executed properly. Additionally, RCM's principles are so diverse that they can be applied to any asset—an airplane, nuclear power plant, truck, tank, ship, manufacturing plant, offshore oil platform, mobile air conditioning unit, tow tractor, jet engine, a single pump, or an engine control unit. RCM principles can be widely applied to an entire asset or more narrowly applied to select pieces of equipment.

Chapter 1

1.1 What Is RCM?

RCM can also be used to formulate scores of solutions that reach far beyond maintenance.

The name *Reliability Centered Maintenance* lends itself to a process that is used to develop proactive maintenance for an asset, but RCM can also be used to formulate scores of solutions that reach far beyond maintenance. These solutions can offer tremendous benefit to an organization. Nevertheless, when applying RCM, many organizations focus only on the development of a proactive maintenance program, which doesn't take full advantage of RCM's powerful principles. This book sets forth the principles of RCM in a straightforward manner so that those interested in applying RCM can be aware of not only how uncomplicated the application of RCM can be, but also how powerful it is.

1.2 Elements that Influence a System

It is especially important to look beyond proactive maintenance because there are so many elements that influence a system, as depicted in Figure 1.1.

It doesn't matter what the equipment is. Many factors have a direct effect on equipment performance: the scheduled main-

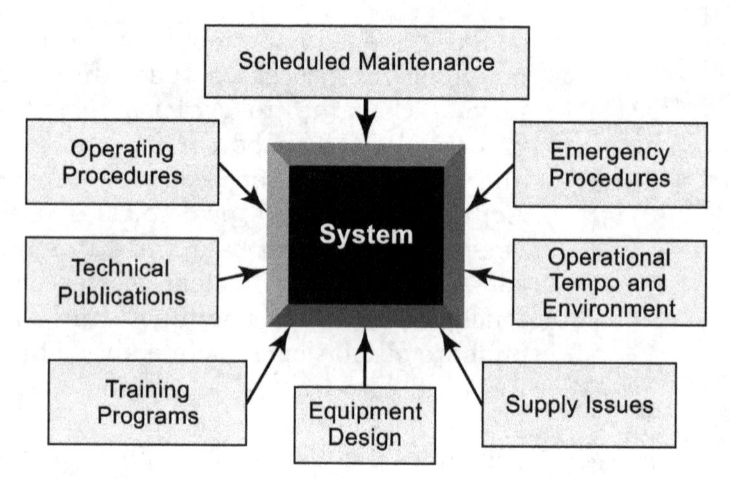

Figure 1.1 Examples of elements that influence a system

tenance that is applied, the operating procedures that are performed, the technical publications that are referenced, the training programs that are attended, the design features that are in service, the spare parts (or lack thereof) that are relied upon, how often an asset is operated, where equipment is required to function, and the emergency procedures that are in place. If these strategies are well developed, the equipment (and thus the organization) benefit. If any of these strategies are ill-conceived or inappropriate, the process by which the equipment plays a part suffers.

1.3 The Essence of RCM: *Managing the Consequences of Failure*

It is often wrongly believed that equipment custodians are in the business of *preventing* failure. Although it is possible to develop strategies that do prevent *some* failures (see Chapter 9), it is nearly impossible to *prevent* all failures. For example, is it possible to prevent *all* failures associated with an electric motor? How about an automobile starter, avionics equipment, or a turbine engine? Certainly not. Thus, other strategies are often put in place in order to manage otherwise *unpreventable* failures when they occur.

Responsible custodians are in the business of managing the consequences of failure—not necessarily preventing them.

For example, organizations rely heavily on operating procedures, emergency procedures, training programs, and redundancy in the design of equipment, as depicted in Figure 1.1. There are three fully redundant hydraulic systems on most commercial aircraft because it is understood that all causes of failure for a hydraulic system cannot be prevented. If one of the three systems fails, two fully redundant systems are available to provide the required hydraulic power for safe flight. Because all failures cannot be prevented, responsible custodians must put other solutions in place to properly deal with failure when it occurs. In other words, responsible custodians are in the business of *managing the consequences of failure*—not necessarily preventing them.

1.4 What RCM Can Yield

Myriad issues, such as incomplete operating procedures or poor equipment design, can negatively affect equipment performance. For that reason, it is incredibly important that these issues are identified and included in an RCM analysis. Including them allows the matter to be analyzed using RCM principles so that a technically appropriate and effective solution can be formulated.

One of the major products of an RCM analysis is the development of a scheduled maintenance program. However, as depicted in Figure 1.2, RCM can help formulate other solutions such as the development of a proactive maintenance plan, new operating procedures, updates to technical publications, modifications to training programs, equipment redesigns, supply changes, enhanced troubleshooting procedures, and revised emergency procedures.

In the context of RCM, these other solutions are referred to as default strategies, as depicted in Figure 1.3.

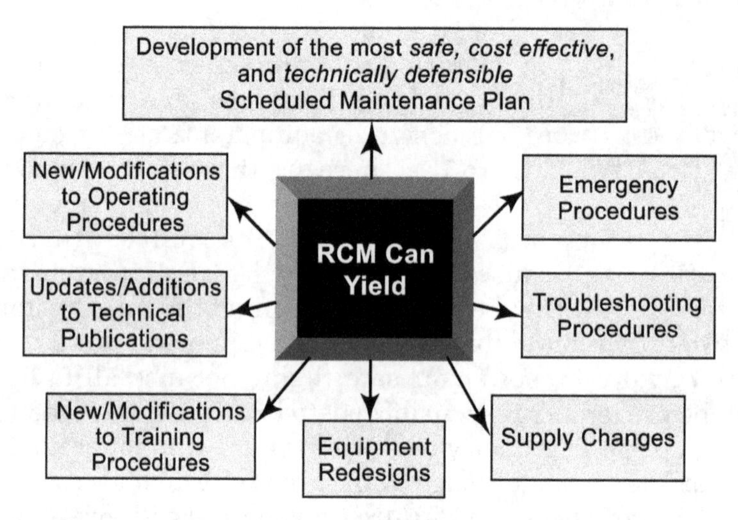

Figure 1.2 Examples of solutions that RCM can yield

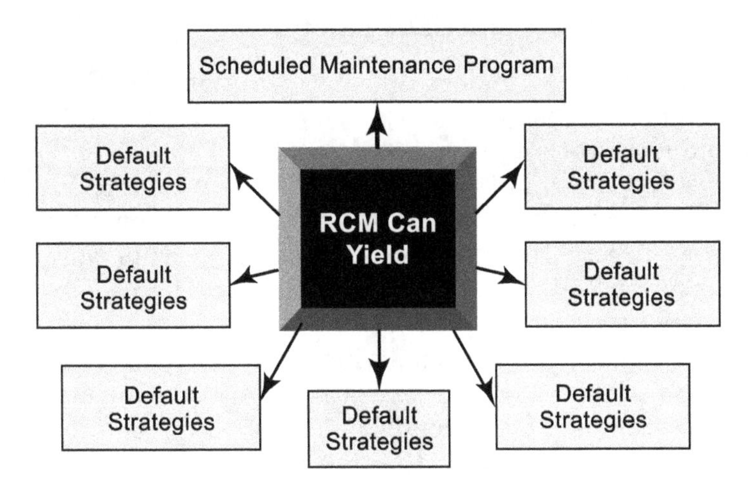

Figure 1.3 RCM can yield a scheduled maintenance program and default strategies

In the context of RCM, together, scheduled maintenance tasks and default strategies are referred to as *failure management strategies*, as depicted in Figure 1.4. These solutions are designed to *manage* failure.

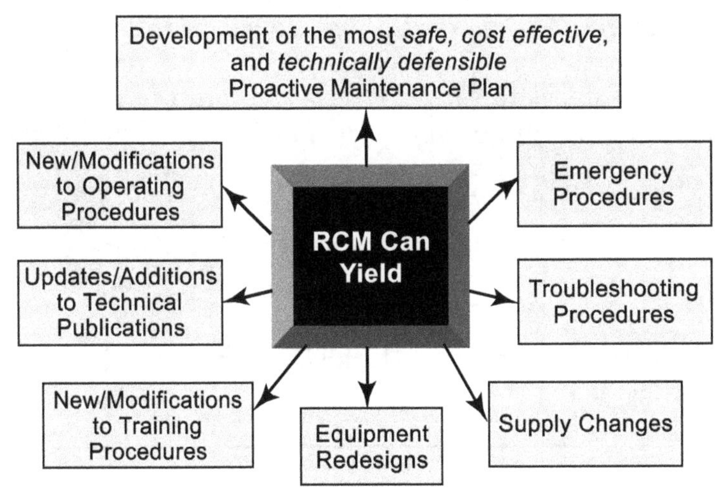

Figure 1.4 Failure management strategies

1.5 The Evolution of RCM Principles

It is important to understand the evolution of RCM in order to appreciate the majesty of its principles. RCM's evolution is best told as a story, as it was told to me.

The story starts in the mid 1950s in the commercial airline industry where, at the time, it was believed that nearly all failures were directly related to operating age. In other words, failure was more likely to occur as operating age increased. Figure 1.5 illustrates this point.

The *x*-axis represents age, which can be measured in any units such as calendar time, operating hours, miles, and cycles. The *y*-axis represents the conditional probability of failure. The philosophy associated with the failure pattern is that, assuming an item stays in service and reaches the end of the useful life, the probability of failure greatly increases if it remains in service. In other words, as stated by United Airlines' Stanley Nowlan and Howard Heap, it was believed that "every item on a complex piece of equipment has a 'right age' at which complete overhaul is necessary to ensure safety and operating reliability." Therefore, it was believed that the sensible thing to do was to overhaul or replace components *before* reaching the end of the useful life with the belief that this would *prevent* failure.

The mindset that failure was more likely to occur as operating time increased was deeply embedded in the maintenance programs. At the time, approximately 85% of aircraft components were subject to fixed interval overhaul or replacement. The maintenance programs were very high in scheduled over-

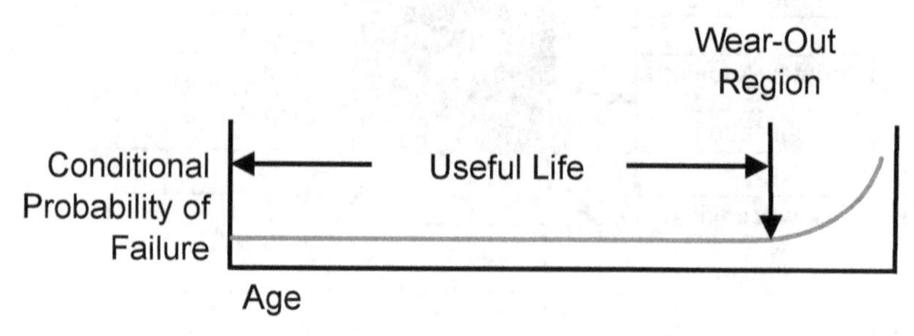

Figure 1.5 Traditional view of failure

hauls and scheduled replacements.

Time marched on. By the late 1950s, new aircraft emerged that included brand new and more technologically advanced equipment such as electronics, hydraulics, pneumatics, pressurized cabins, and turboprop engines. Because the equipment was new, there was no operational experience or any historical failure data available. Therefore, the useful life of the new equipment components was unknown. However, a maintenance plan still had to be developed. As a result, the new plans were mirrored from the old plans. For the new equipment where there was no current maintenance to mirror, they took their best educated guess. The aircraft were sent into service and maintained using maintenance plans formulated in this manner.

By the early 1960s, failure data had been accumulated. Worldwide, the crash-rate was greater than 60 crashes per million takeoffs, and two-thirds of these crashes were due to equipment failure. To put this crash rate into perspective, that same crash rate in 1985 would be the equivalent of two Boeing 737s crashing somewhere in the world every day.

The increased crash rate became an issue for operations, management, government, and regulators, so action was taken in an attempt to increase equipment reliability. Consistent with the philosophy at the time—that failure was directly related to operating age (as depicted in Figure 1.6)—the overhaul and replacement intervals were shortened, thereby *increasing* the amount of maintenance that was performed and *increasing* maintenance downtime. An example of a shortened overhaul inteval is depicted in Figure 1.6.

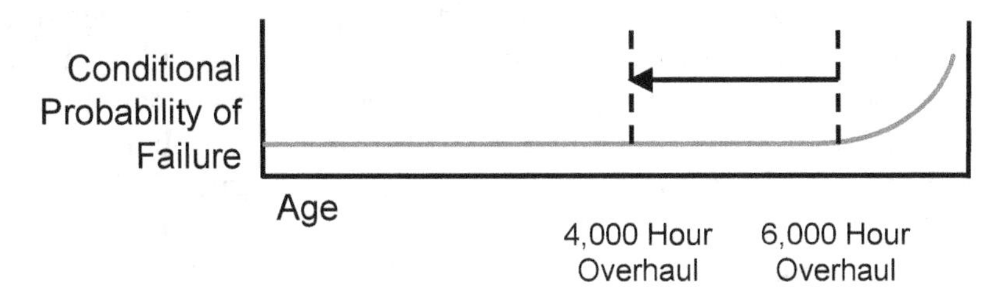

Figure 1.6 Example of a shortened overhaul interval

The new maintenance plans were put into service. After a period of time, they noticed that three things happened.

1. In very few cases things got better.
2. In very few cases things stayed the same.
3. But, for the most part *things got worse.*

The Federal Aviation Administration (FAA) and industry were frustrated by their inability to control the failure rate by changing the scheduled overhaul and replacement intervals. As a result, a task force was formed in the early 1960s. This team of pioneers was charged with the responsibility of obtaining a better understanding of the relationship between operating reliability and policy for overhaul and replacement.

They identified that two assumptions were embedded in the current maintenance philosophy.

Assumption 1: The likelihood of failure increases as operating age increases.
Assumption 2: It is assumed we know when those failures will occur.

The team identified that the second assumption had already been challenged. In an attempt to decrease the failure rate, the overhaul and replacement intervals were shortened, as depicted in Figure 1.6. But when the intervals were shortened, the failure rate increased. It was then identified that the first assumption—the likelihood of failure increases as operating age increases—needed to be challenged.

As a result, an enormous amount of research was performed. Electronics, hydraulics, pneumatics, engines, and structures were analyzed. What was discovered rocked the world of maintenance at the time. The research showed that there wasn't one failure pattern that described how Failure Modes behave. In fact there are six failure patterns, as seen in Figure 1.7.

Failure patterns A, B, and C all have something in common. They exhibit an age-related failure phenomenon. Likewise, failure patterns D, E, and F have something in common. They exhibit randomness.

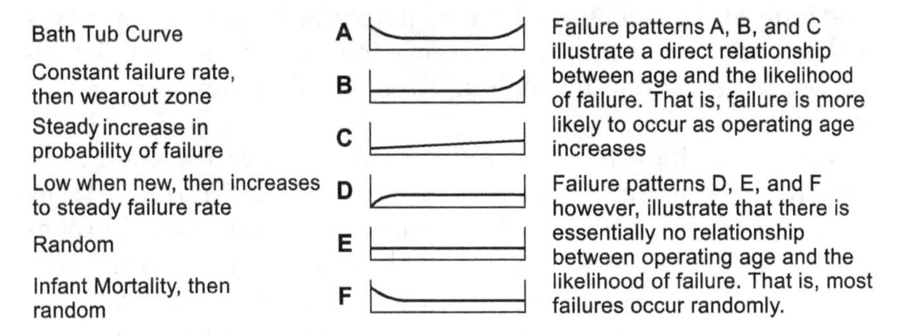

Figure 1.7 Six patterns of failure

What was especially shocking was the percentage of Failure Modes that conformed to each failure pattern. Figure 1.8 summarizes the percentage of Failure Modes conforming to each failure pattern.

Bath Tub Curve	A	4%
Constant failure rate, then wearout zone	B	2%
Steady increase in probability of failure	C	5%
Low when new, then increases to steady failure rate	D	7%
Random	E	14%
Infant Mortality, then random	F	68%

Figure 1.8 Percentages of Failure Modes that conformed to each failure pattern

Collectively, only 11 percent of aircraft system Failure Modes behaved according to failure patterns A, B, and C, where the likelihood of failure rises with increased operating age. Failure patterns A and B have a well-defined wearout zone; it makes sense that Failure Modes conforming to these failure patterns could effectively be managed with a fixed interval overhaul or replacement. Failure patterns A, B, and C are typically associated with simple items that are subject to, for example, fatigue or wear such as tires, brake pads, and aircraft structure.

However, the remaining 89 percent of aircraft system Failure Modes occur randomly. They correspond to failure patterns D, E, and F. After the short increase in the conditional probability of failure in pattern D, as well as the infant mortality period present in failure pattern F, the Failure Mode has the same likelihood of occurring at any interval in the equipment's expected service life. Therefore, for 89% of Failure Modes, it makes no sense to perform a fixed interval overhaul or replacement because the probability of failure is constant. These failure patterns are typically associated with complex equipment such as electronics, hydraulics, and pneumatics.

Two most notable issues

1. Only two percent of the Failure Modes conformed to failure pattern B as shown in Figure 1.9, yet this was the failure pattern that defined the way they believed equipment failure behaved!

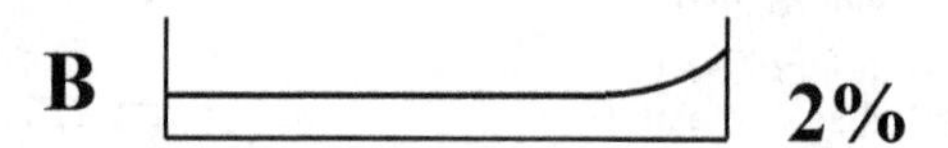

Figure 1.9 Percentage of Failure Modes that conformed to Failure Pattern B

2. After the short increase in the conditional probability of failure in pattern D, as well as the infant mortality period present in failure pattern F, 89 percent of Failure Modes occur randomly, as depicted in Figure 1.10.

What was astonishing was that the maintenance plans in use were derived assuming nearly all Failure Modes behaved according to failure pattern B. Yet only two percent of the Fail-

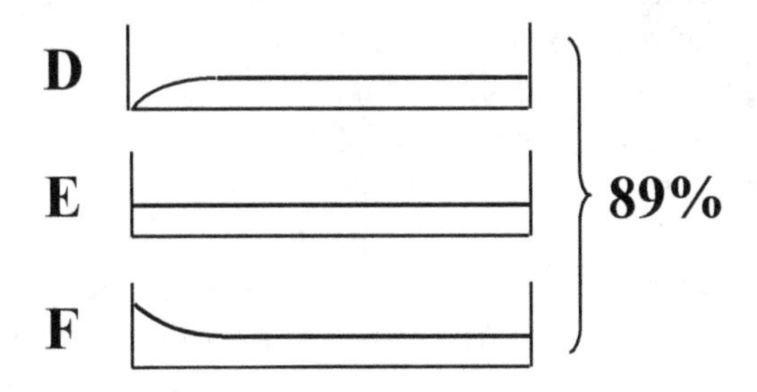

Figure 1.10 Percentage of Failure Modes that conformed to Failure Patterns D, E, and F

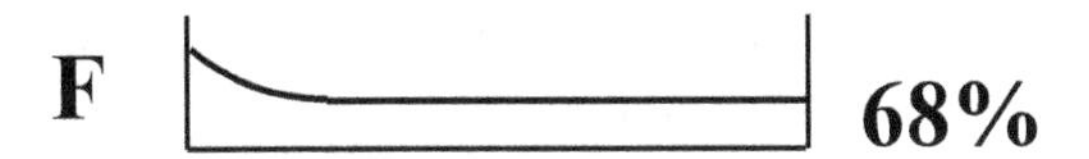

Figure 1.11 Percentage of Failure Modes that conformed to Failure Pattern F

ure Modes actually behaved that way. Furthermore, it was shown that most Failure Modes occur randomly. Therefore, fixed interval overhaul or replacement technically made no sense. That is, if an item is replaced today, it has the same chance of failing tomorrow as it does one year later.

More important, not only were the vast majority of scheduled overhauls and replacements senseless, their efforts to control the failure rate with fixed interval overhaul and replacement were counterproductive. Their study showed that 68 percent of Failure Modes behaved according to failure pattern F, as depicted in Figure 1.11.

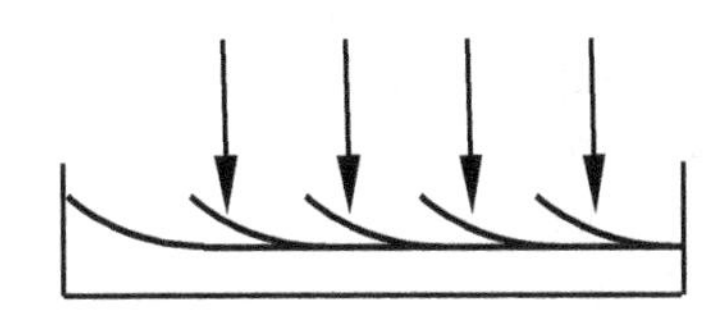

Figure 1.12 Reintroducing infant mortality

Infant mortality (e.g., component installed backwards, tool left behind, poor operating procedures) played a significant role in the high unreliability rates. Therefore, these weaknesses were making things worse with scheduled overhauls and replacements. As depicted in Figure 1.12, each time a scheduled overhaul or replacement was performed, infant mortality was reintroduced into an otherwise stable system.

Because most Failure Modes occur randomly, the failure rate could not be controlled by performing more scheduled overhauls and replacements.

This research conclusively proved that fixed interval overhaul or replacement is technically not the right action to take when failure is not a function of operating age. In fact, in most cases, scheduled overhaul and replacement hurt reliability. Because most Failure Modes occur randomly, the failure rate could not be controlled by performing more scheduled overhauls and replacements. Armed with these facts, a new way of deriving scheduled maintenance tasks needed to be developed, setting the stage for the birth of RCM principles.

1.6 The Development of RCM Principles

From this research, RCM principles were first conceived within the commercial airline industry. *MSG-1, Handbook: Maintenance Evaluation and Program Development* was prepared by the 747 Maintenance Steering Group and published in 1968. This document contained the first use of decision diagram techniques to develop a prior-to-service scheduled maintenance program.

Improvements to MSG-1 led to the development of *MSG-2: Airline/Manufacturer Maintenance Program Planning Document,* which was published in 1970. MSG-2 was used to develop the scheduled maintenance programs for the Lockheed 1011 and the Douglas DC-10. It was also used on tactical military aircraft McDonnell F4J and the Lockheed P-3.

In the mid-1970s, the Department of Defense was interested in learning more about how maintenance plans were developed within the commercial airline industry. In 1976 the Department of Defense commissioned United Airlines to write a report that detailed their process. Stanley Nowlan and Howard Heap, engineers at United Airlines, wrote a book on the process and called it *Reliability-Centered Maintenance.* Their book was published in 1978. To many, Stanley Nowlan and Howard Heap are considered two of the most significant pioneers of the RCM process. Their book remains one of the most important documents ever written on equipment maintenance.

Using Nowlan and Heap's book as a basis for update, *MSG-3, Operator / Manufacturer Scheduled Maintenance Development* was published in 1980. Since then, MSG-3 has gone through many updates. MSG-3 continues to be used within the commercial airline industry today, but is still intended to develop a scheduled maintenance program for *prior to service* aircraft.

Since Nowlan and Heap's book was published, there have been various updates to the RCM process, namely the identification of environmental issues. The late John Moubray was another great pioneer of the RCM process; he did a great deal to advance RCM throughout commercial industry. His book *RCM II* was first published in the United Kingdom in 1991 and in the United States in 1992.

Streamlined RCM and SAE JA1011

Although RCM is a resource intensive process, analyses can be completed efficiently if the process is used correctly with the right people. However, in the mid 1990s, streamlined versions of RCM started to appear. These versions often omit key steps in the process and differ significantly from what Nowlan and Heap originally intended. As a result, the Society of Automotive Engineers (SAE) published *SAE JA1011*, Evaluation *Criteria for Reliability-Centered Maintenance (RCM) Processes* in 1999. This internationally-recognized standard outlines the criteria that any RCM process must embody in order to be called RCM. SAE JA1011 was updated in 2009.

The RCM process defined in this book complies with SAE JA1011. More important, it remains true to what the original pioneers of the process, Stanley Nowlan and Howard Heap, originally intended. Therefore, this books details *True RCM.*

1.7 Definition of RCM

RCM is a remarkable process and can be defined as follows. The terms *zero based, failure management strategies,* and *operational environment* bear further explanation.

Reliability Centered Maintenance is a **zero-based,** structured process used to identify the **failure management strategies** required to ensure an asset meets its mission requirements in its **operational environment** in the *most safe* and *cost effective* manner.

Zero-based

Each RCM analysis is carried out assuming that no proactive maintenance is being performed. In other words, Failure Modes and Failure Effects are written assuming that nothing is being done to predict or prevent the Failure Mode. In this way, consequences of each Failure Mode can be assessed and solutions can be formulated with no bias towards what is currently being done.

Failure Management Strategies

Notice that the definition states that RCM is used to identify *failure management strategies*, not *maintenance tasks*. As explained earlier, managing assets requires more than just scheduled maintenance. Therefore, RCM provides powerful tools for developing other solutions, as detailed in Figure 1.2.

Operational Environment

How an asset is maintained depends on far more that just what an asset is. When solutions for assets are formulated, the following issues regarding the operational environment must be considered.

- Physical environment in which the asset will be used (e.g., cold weather, desert climate, controlled environment)
- Operational tempo (e.g., 24 hour operation, system runs 6 hours each day)

- Circumstances under which the system will be operated (e.g., stand-alone, one of four systems runs at one time but is rotated every month)
- Redundancy (e.g., the system or any of its components operate in the presence of a backup)

These issues can greatly influence not only what maintenance tasks are identified and how often they are performed, but also other solutions such as equipment design and training programs. Therefore, the operational environment must be clearly defined.

1.8 Defining Performance in the Context of RCM

In the context of RCM, there are two features regarding equipment performance that responsible custodians must carefully examine: *design capability* and *required performance.*

When it comes to defining performance, equipment custodians must be specific about what their assets can do (design capability) and what they need them to do (required performance).

Asset owners perform RCM to determine what actions must be taken to ensure that equipment meets mission requirements. A mission could be towing a piece of equipment to the construction site, launching an aircraft from an aircraft carrier, or ensuring that there is adequate plant air for the downstream manufacturing process. But when it comes to defining performance, equipment custodians must be specific about what their assets can do (design capability) and what they need them to do (required performance). The following discussion illustrates this point.

Take, for example, a water tube steam boiler. As illustrated in Figure 1.13, the design capability is a Maximum Allowable Working Pressure (MAWP) of 500 psi. However, the required performance is 650 psi. Is this scenario acceptable? Absolutely not, because what the organization requires (650 psi) exceeds the design capability of the boiler (500 psi).

Figure 1.14 illustrates another example. Here, the design capability is an MAWP of 650 psi and the required perform-

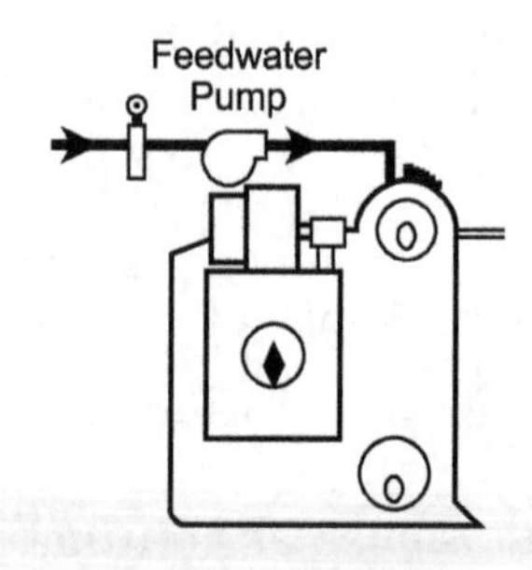

Water Tube Steam Boiler

Design Capability MAWP 500 psi	Required Performance 650 psi

Figure 1.13 Organizational requirements exceed design capability

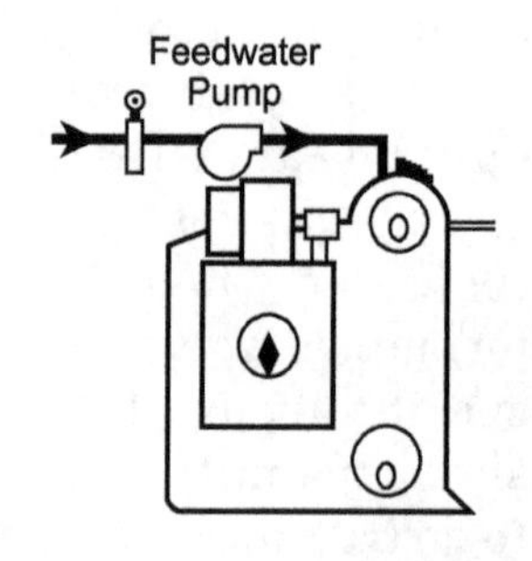

Water Tube Steam Boiler

Design Capability MAWP 650 psi	Required Performance 500psi

Figure 1.14 Organizational requirements fit within the design capability of the asset

ance is 500 psi. Is this scenario acceptable? Yes, because what the organization requires (500 psi) fits within the design capability of the asset.

This may seem to be an incredibly simple concept—so basic and fundamental that it doesn't even warrant being mentioned. It appears that way. However, this concept is a very serious issue. If an organization gets it wrong, it can turn deadly. In fact, it has turned deadly.

Three Air Tanker Crashes

The National Transportation Safety Board (NTSB) investigated three air tanker crashes. The following information was reported in the NTSB Safety Recommendation dated April 23, 2004.

> On August 13, 1994, a Lockheed C-130A Hercules experienced an in-flight separation of the right wing near Pearblossom, California, while responding to a forest fire near the Tahachapi Mountains. All three crewmembers were killed and the airplane was completely destroyed. (An aircraft similar to the C-130A can be seen in Figure 1.15.)

***Figure 1.15 C-130 Aircraft, similar to the C-130A Tanker that crashed on August 13, 1994 and June 17, 2002** (Photo from Photo NSA online; http://www.nsa.gov/about/photo_gallery/index.shtml.)*

***Figure 1.16 C130A June 17, 2002, crash site from the NTSB report** (Photo from NTSB, September 24, 2002, NTSB Advisory, Update on Investigations of Firefighting Airplane Crashes in Walker, California and Estes Park, Colorado; http://www.ntsb.gov/pressrel/2002/020924.htm)*

On June 17, 2002, another Lockheed C-130A Hercules experienced an in-flight breakup that was initiated by separation of the right wing, followed by separation of the left wing, while executing a fire retardant drop over a forest fire near Walker, California. Both wings detached from the fuselage at their respective center wing box-to-fuselage attachment locations. All three flight crewmembers were killed, and the airplane was completely destroyed. Figure 1.16 depicts the June 2002 crash site.

On July 18, 2002, a Consolidated Vultee P4Y Privateer experienced an in-flight separation of the left wing while maneuvering to deliver fire retardant over a forest fire near Estes Park, Colorado. Both crewmembers were killed and the airplane was destroyed. (A similar aircraft is shown in Figure 1.17.)

Figure 1.17 P4Y Privateer similar to the one that crashed on July 18, 2002 (Library of Congress, Prints & Photographs Division, FSA/OWI Collection, [LC-USE6- D-009930])

All three aircraft were leased by the U.S. Department of Agriculture's Forest Service for public firefighting flights. However, the aircraft detailed above were originally designed to transport cargo for the U.S. military—*not to fight forest fires.*

Air Tanker Crashes: Design Capability versus Required Performance

The operational environment and the loads experienced by an aircraft transporting cargo are vastly different from those experienced by an aircraft fighting forest fires. The NTSB report explains that during a fire-fighting mission, an aircraft experiences "frequent and aggressive low-level maneuvers with high acceleration loads and high levels of atmospheric turbulence." The NTSB report further details that the maintenance programs used for the aircraft were the same that were derived for the aircraft when their mission was transporting cargo for the military. The report states that the aircraft were likely "operating outside the manufacturers' original design intent."

In the context of RCM, the *required performance* of the organization using the air tankers far exceeded the *design capability* of the aircraft. The structural lives of the aircraft were shortened because of the harsh operating environment and the far more aggressive loads applied to the aircraft during firefighting versus transporting cargo. The increased loading accelerated fatigue crack initiation and sped up the crack propagation time. Therefore, the structural inspections that were in place were not accomplished often enough to identify the crack before it caused catastrophic failure. The simple concept of ensuring that an asset is capable of doing what the organization requires was completely overlooked.

Aloha Airlines, Flight 243

On April 28, 1988, Aloha Airlines, Flight 243 took off from Hilo, Hawaii, at 1:25 p.m. Shortly after the aircraft leveled off at 24,000 feet, the aircraft experienced explosive decompression and structural failure that ripped away a large section of the fuselage, as shown in Figure 1.18. One of the flight attendants, Clarabelle Lansing was immediately wrenched from the airplane. The aircraft made an emergency landing at Kahului Airport. The 89 passengers onboard and the remaining 4 crewmembers survived.

This tragedy is another example of required performance being allowed to exceed design capability.

Aloha Airlines: Design Capability versus Required Performance

Aloha Airlines was using its 737s for inter-island Hawaiian flights. According to the NTSB Aircraft Accident Report, those aircraft were accumulating three flight cycles (take-off and landing) for every hour in service. However, Boeing designed the structural inspections for the 737 assuming that the aircraft would accumulate about one and a half cycles per flight hour. Therefore, the aircraft were accumulating flight cycles at twice the rate for which the Boeing Maintenance Planning Data (MPD) was designed. Similar to the air tanker crashes described previously, this use accelerated fatigue crack initiation and increased the crack propagation time. The structural inspections and associated intervals in place were inadequate;

Figure 1.18 Aloha Airlines, Flight 243, April 28, 1988 after landing (Associated Press / Robert Nichols)

they were not accomplished frequently enough to detect the crack before catastrophic failure occurred.

The air tanker fatal crashes and the Aloha Airlines' accident are only two examples that underscore the critical importance of ensuring that an asset's design capability is capable of meeting organizational requirements. It is a simple concept that is too often overlooked. During an RCM analysis, asset design capability and required performance are carefully analyzed.

1.9 Introduction to the RCM Process

The application of True RCM consists of preparing an Operating Context and carrying out the 7 steps of RCM.

The application of *True RCM* consists of preparing an Operating Context and carrying out the 7 steps of RCM. Figure 1.19 outlines the RCM process.

Chapters 2 through 8 detail the Operating Context and the seven steps of the RCM process. The following discussion briefly introduces each concept.

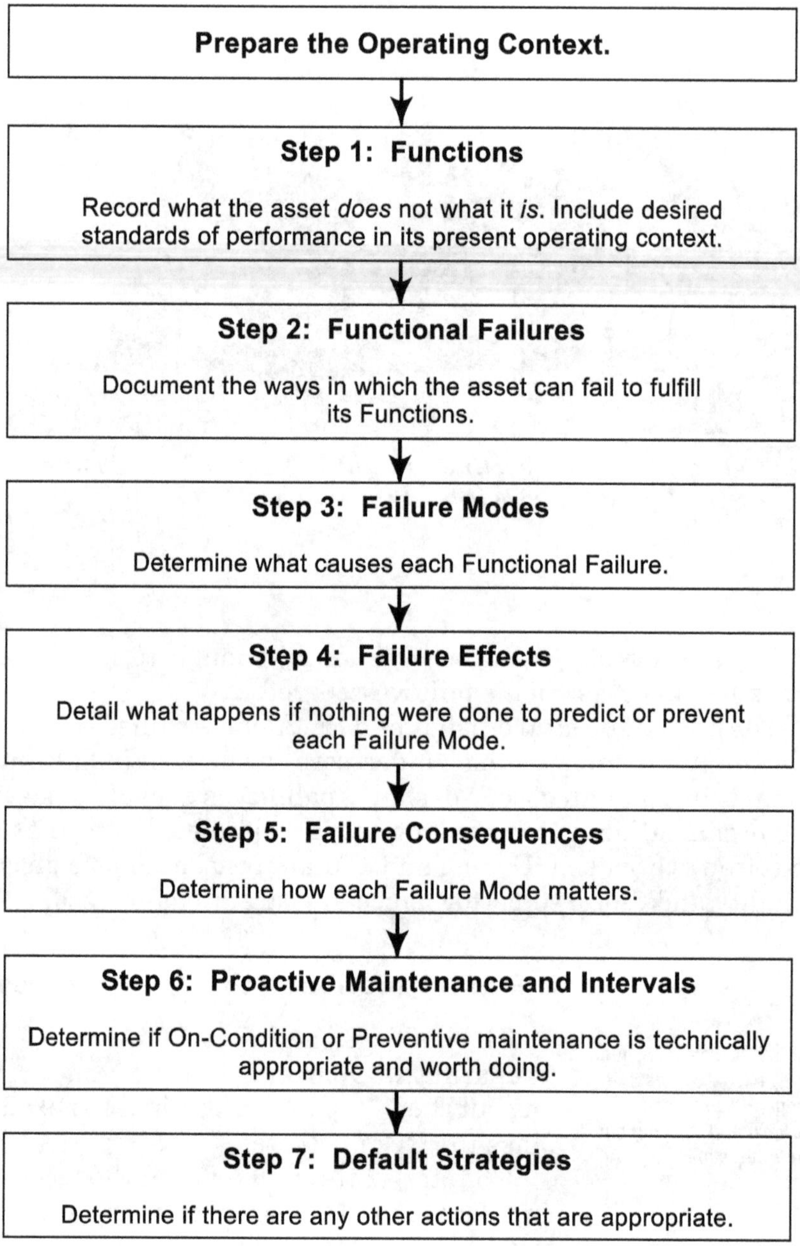

Prepare the Operating Context.

↓

Step 1: Functions

Record what the asset *does* not what it *is*. Include desired standards of performance in its present operating context.

↓

Step 2: Functional Failures

Document the ways in which the asset can fail to fulfill its Functions.

↓

Step 3: Failure Modes

Determine what causes each Functional Failure.

↓

Step 4: Failure Effects

Detail what happens if nothing were done to predict or prevent each Failure Mode.

↓

Step 5: Failure Consequences

Determine how each Failure Mode matters.

↓

Step 6: Proactive Maintenance and Intervals

Determine if On-Condition or Preventive maintenance is technically appropriate and worth doing.

↓

Step 7: Default Strategies

Determine if there are any other actions that are appropriate.

Figure 1.19 The RCM Process

Operating Context

An Operating Context is a document that includes relevant technical information such as the scope of analysis, theory of operation, equipment description, and RCM analysis notes. In essence, it is a storybook identification of the system to be analyzed. The Operating Context also documents notes and assumptions regarding analysis decisions. It is an important source of reference for working group and validation team members.

In the interest of time, the Operating Context is typically drafted by the facilitator before the analysis begins and is then reviewed with the working group before the first step in the RCM process (identifying Functions) is accomplished. During this time, the working group reviews and revises the Operating Context, as required. The Operating Context is considered a living document; it is edited as more is learned about the equipment and additional issues come to light during the analysis.

Step 1: Functions

The intention of RCM is to determine what solutions must be put in place to ensure an asset meets the requirements of the organization. The air tanker crashes and the Aloha Airlines disaster detailed previously illustrate how critical it is to understand what is *required* of an asset so that it can be determined if the asset is capable of fulfilling those requirements. For this reason, the first step in the RCM process is to identify Functions.

Functions and associated performance standards are always written to reflect what the *organization requires* from the asset rather than what the system is designed to provide. During Function development, it is often noted that the organization's expectations of the equipment exceed the actual capabilities of the asset. As depicted in Figure 1.20, the Primary Function (the main purpose the system exists) and Secondary Functions (other Functions of the asset) are recorded.

Step 2: Functional Failures

Step 2 in the RCM process is to identify Functional Failures for each Function. Nowlan and Heap define Functional Failure as an *unsatisfactory condition*. As depicted in Figure 1.21, both

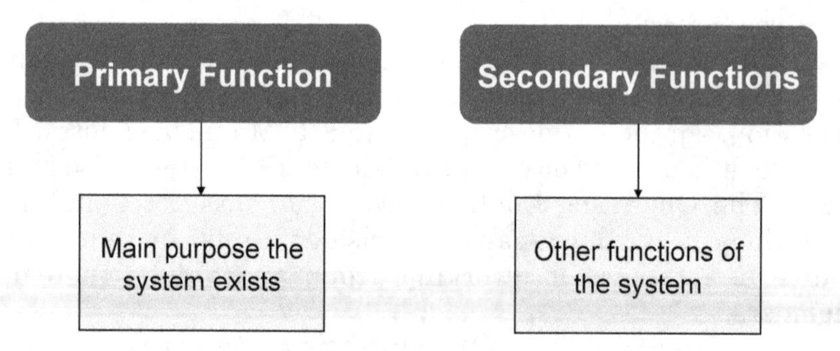

Figure 1.20 Primary and Secondary Functions

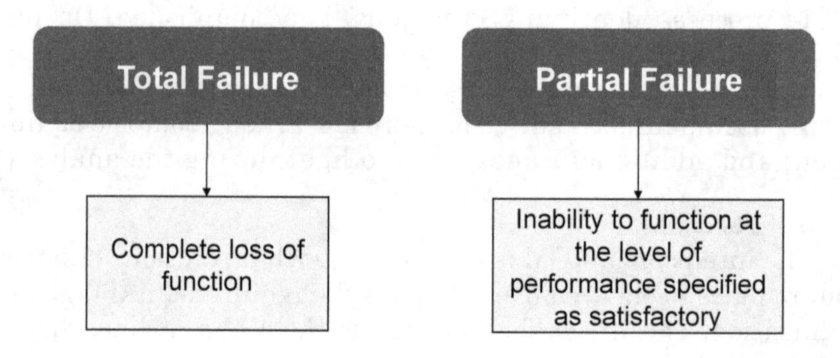

Figure 1.21 Total and Partial Failures

Total and Partial Functional Failures are recorded for each Function. A Total Failure means no part of that Function can be performed. Partial Failure describes how the Function is still possible but is performed at an unsatisfactory level.

Step 3: Failure Modes

A Failure Mode is *what causes a Functional Failure.* During Step 3 of the RCM process, Failure Modes that cause each Functional Failure are identified. It is often wrongly believed that *all* Failure Modes associated with the system being analyzed must be recorded. On the contrary, RCM provides specific guidelines for determining what Failure Modes to include in an analysis. Only Failure Modes that are *reasonably likely to occur* in the operating context should be included. If the answer to one or more of the following questions is "*yes*," the Failure Mode should be included in the analysis:

- Has the Failure Mode *happened before*?
- If the Failure Mode *has not happened*, is it *a real possibility*?
- Is the Failure Mode *unlikely to occur* but the *consequences are severe*?
- Is the Failure Mode *currently managed via proactive maintenance*?

Failure Modes included in most analyses consist of typical causes such as those due to wear, erosion, corrosion, etc. However, it is very important to include Failure Modes that cover issues such as human error, incorrect technical manuals, inadequate equipment design, and lack of emergency procedures. Such Failure Modes allow issues to be analyzed as part of the RCM process so that solutions in addition to proactive maintenance can be developed.

Step 4: Failure Effects

During Step 4, a Failure Effect is written for each Failure Mode. A Failure Effect is a brief description of what would happen if nothing were done to predict or prevent the Failure Mode. Failure Effects should be written in enough detail so that the next step in the RCM process, Failure Consequences, can be identified. Failure Effects should include:

- Description of the failure process from the occurrence of the Failure Mode to the Functional Failure
- Physical evidence that the failure has occurred
- How the occurrence of the Failure Mode adversely affects safety and/or the environment
- How the occurrence of the Failure Mode affects operational capability/mission

- Specific operating restrictions as a result of the Failure Mode
- Secondary damage
- What repair is required and how long it is expected to take

Information Worksheet

Steps 1 through 4 of the RCM process are recorded in the *Information Worksheet*, as depicted in Figure 1.22. The Information Worksheet includes Functions, Functional Failures, Failure Modes, and Failure Effects.

Information Worksheet

	Function		Functional Failure		Failure Mode	Failure Effect
1		A		1		
1		A		2		
1		B		1		

Figure 1.22 The Information Worksheet

Step 5: Failure Consequences

A properly written Failure Effect allows the Failure Consequence to be assessed. A Failure Consequence describes how the loss of function caused by the Failure Mode matters. There are four categories of Failure Consequences:

- Safety
- Environmental
- Operational
- Non-Operational

Step 6: Proactive Maintenance and Associated Intervals

After consequences are assessed, the next step in the RCM process is to consider proactive maintenance as a failure management strategy. In the context of RCM, the proactive maintenance tasks that may be identified include:

Scheduled Restoration A scheduled restoration task is performed at a specified interval to restore an item's failure resistance to an acceptable level—without considering the item's condition at the time of the task. An example of a scheduled restoration task is retreading a tire at 60,000 miles.

Scheduled Replacement A scheduled replacement task is performed at a specified interval to replace an item without considering the item's condition at the time of the task. An example is a scheduled replacement of a turbine engine compressor disk at 10,000 hours.

Scheduled restorations and scheduled replacement tasks are performed at specified intervals regardless of the item's condition.

On-Condition Task An On-Condition task is performed to detect evidence that a failure is impending. In the context of RCM, the evidence is called a *potential failure condition* and can include increased vibration, increased heat, excessive noise, wear, etc. Potential failure conditions can be detected using relatively simple techniques such as monitoring gauges or measuring brake linings. Additionally, potential failure conditions can be detected by employing more technically involved techniques such as thermography or eddy current, or by using continuous monitoring with devices such as strain gauges and accelerometers installed directly on machinery. The point of On-Condition tasks is that maintenance is performed only upon evidence of need.

In the context of RCM, all proactive maintenance tasks must be technically appropriate and worth doing. Chapter 9 details how to determine if a proactive task is technically appropriate and worth doing.

Step 7: Default Strategies

As mentioned earlier, RCM isn't just about maintenance. There are a great many solutions other than proactive maintenance that can be derived using the RCM process. Examples include: Failure Finding tasks, Procedural Checks, no scheduled maintenance, and other recommendations such as modifications to operating procedures, updates to technical publications, and equipment redesigns. In the context of RCM, these recommendations are known as Default Strategies. Default Strategies are discussed in detail in Chapter 10.

1.10 Failure Modes and Effects Analysis (FMEA) and Failure Modes, Effects, and Criticality Analysis (FMECA)

It is often wrongly believed that FMEA and FMECA are analyses that are accomplished independently of, or in lieu of, RCM. On the contrary, the first four steps of the RCM process produce a FMEA. The steps to accomplish a FMEA are depicted in Figure 1.23.

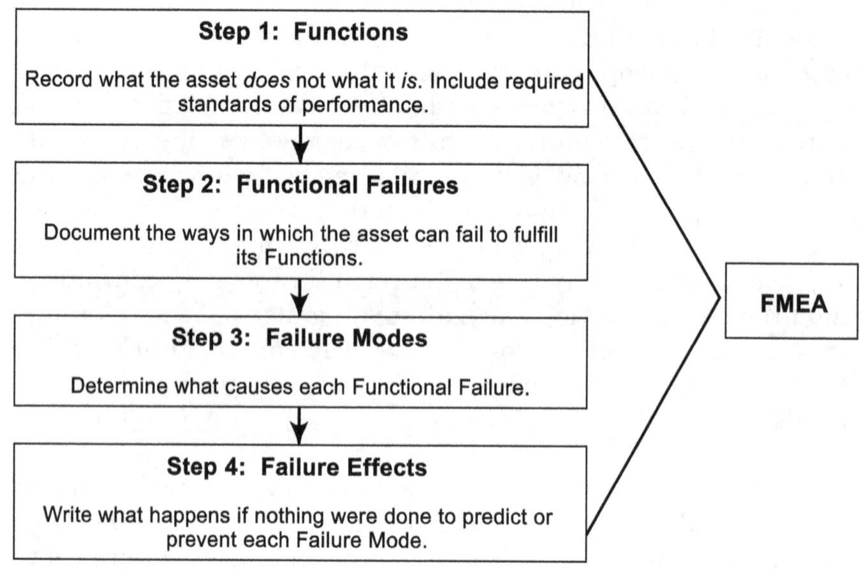

Figure 1.23 First four steps of the RCM process produce a FMEA

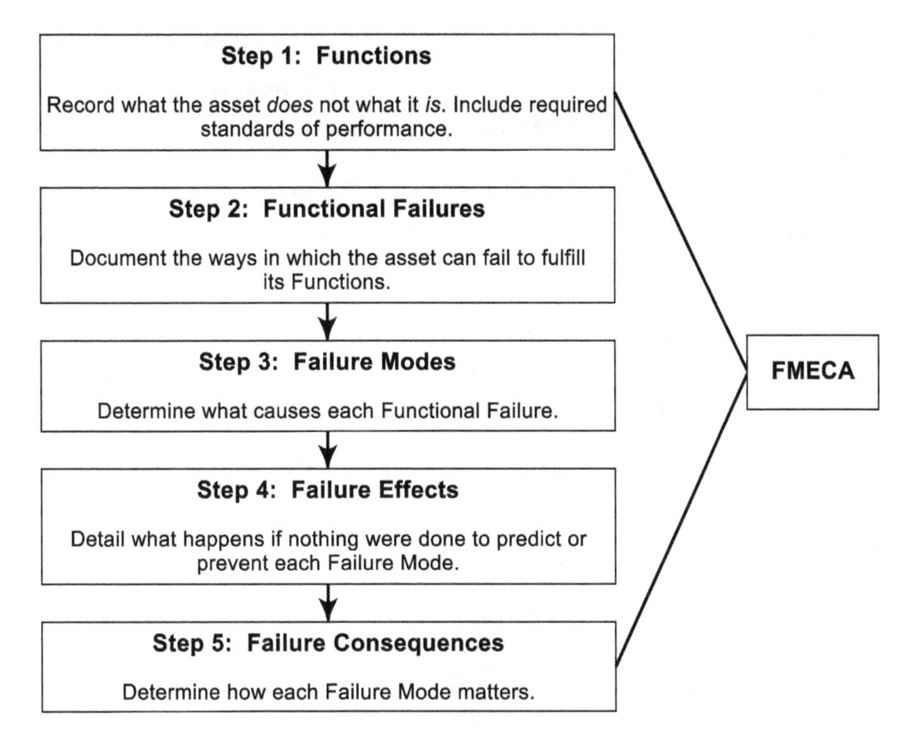

Figure 1.24 First five steps of the RCM process produce a FMECA

Additionally, the first five steps of the RCM process generate a FMECA. The steps to accomplish a FMECA are depicted in Figure 1.24.

When RCM is performed, the requirement for a FMEA and a FMECA is largely satisfied.

Chapter 1

Summary

RCM is an exciting process that yields overwhelming positive results when the process is applied correctly with the right people. RCM isn't a new process. The application of its principles spans several decades and has been (and is being) applied in nearly every industry throughout the world. RCM can be carried out swiftly and efficiently when executed properly. Additionally, RCM's principles are so diverse that they can be applied to any asset such as an airplane, nuclear power plant, manufacturing plant, or an offshore oil platform. RCM principles can be widely applied to an entire asset or more narrowly applied to select pieces of equipment.

After the operating context is drafted, the seven steps of the RCM process are carried out: 1) Functions; 2) Functional Failures; 3) Failure Modes; 4) Failure Effects; 5) Failure Consequences; 6) Proactive Maintenance and Intervals; and 7) Default Strategies. One of the major products of an RCM analysis is the development of a scheduled maintenance program. However, RCM can be used to formulate scores of solutions that reach far beyond maintenance.

A Facilitated Working Group Approach to RCM

When Thomas Edison was asked why he had a team of twenty-one assistants he said:

"If I could solve all the problems myself, I would."

Overwhelming positive results are reaped when equipment experts are empowered to make decisions for physical assets.

It is very exciting to see the overwhelming positive results that are reaped when equipment experts—*those who are intimate with the asset and the operating environment*—are empowered to make decisions for physical assets. In fact, it is such a powerful concept that it is bewildering why organizations don't employ teams more proactively when it comes to asset management.

Still, most organizations today use a single-analyst approach to RCM. That is, an RCM engineer, or in some cases an outside contractor, gathers technical manuals, drawings, etc., and completes the analysis independently. However, this limited perspective typically diminishes the quality and power of the results. In other cases, organizations claim the use of a working group approach by conducting interviews with equipment experts to fill the gaps in a single analyst's analysis. These approaches, which sometimes can become counterproductive, pale in comparison to the remarkable solutions that can be formulated by a team.

2.1 The Team Approach to Accomplishing Objectives

Let's take a look at the team approach because it is essential to the success of so many endeavors. Teams are all around us. For example, one player doesn't win a World Series for a baseball team—nine members are essential to every inning played, and those nine are part of an even larger team. Anyone who has had surgery knows first-hand that there is never just a surgeon in an operating room. Many other professionals are required to ensure a successful procedure—the anesthesiologist, the circulator, the scrub tech, and the first assistant amongst them. How about flying? Is it just a pilot who delivers passengers safely to a destination? Of course not. Many individuals including the copilot, flight attendants, maintenance personnel, ground crew, and air traffic controllers are essential to the flight. In all of these examples, people working together reach a defined purpose.

When there is an aircraft crash in the United States, the National Transportation and Safety Board (NTSB) immediately dispatches a "Go Team." This team can consist of several people up to dozens of individuals representing a variety of disciplines including operations, structures, power plants, systems, weather, and air traffic control. Why are so many people involved in a crash investigation? Because it takes more than one expertise to identify the cause of an aircraft crash. Why then, when it comes to RCM—*a process used to make vital decisions about assets*—would an organization choose to employ a single analyst approach?

Teamwork and Preparation

I had the privilege of hearing Captain Al Haynes, pilot of United Airlines Flight 232, speak at the Aging Aircraft Conference in Missouri in May 2009. United Airlines Flight 232, a DC-10, crashed in Sioux City, Iowa, on July 19, 1989. On that flight, the #2 engine located on the tail of the aircraft suffered an internal engine failure due to an undetected manufacturing defect in the stage one fan rotor assembly. Shrapnel from the failure severed lines in all three, fully redundant hydraulic systems rendering them all completely inoperable. This almost completely crippled the aircraft. All that was left to control the aircraft was the use of the throttles on the #1 and #3 engines. The crew managed to get the aircraft on the ground. Tragically, of the 296 people on board, 112 died—but 184 people lived. Captain Haynes said it was a team of people who allowed so many to live: the airport authorities who readied the airport and runway to accept them, the cabin crew who prepared the passengers for an emergency landing, the air traffic controllers who calmly controlled the aircraft to the runway, the cockpit crew who so expertly managed to get the aircraft on the ground, and emergency personnel who tended to injured passengers. Teamwork!

Another reason, Captain Haynes said that so many people lived was preparation. They were prepared to handle an emergency. He closed his presentation by urging the audience to consider the things that will probably never happen and to prepare for them. His words were, "Be as prepared as you possibly can."

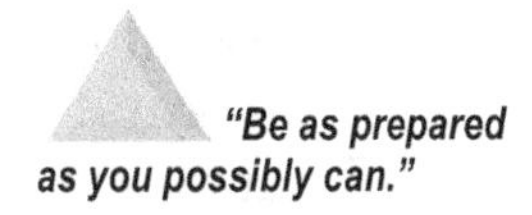

"Be as prepared as you possibly can." These are powerful words when it comes to asset management—especially when considering the types of assets custodians are responsible for and the communities they serve. Organizations need to be prepared to meet mission requirements, production commitments, scheduling constraints, safety goals, environmental regulations, cutbacks of all kinds, quality goals, and cost commitments. Therefore, assets must perform as *required*.

If an organization seeks to be as prepared as it possibly can, who is in the best position to identify and accomplish what that takes? Is it an outside contractor? The equipment manufacturer? The systems engineer? The operator? The maintainer? What one person knows it all? In most cases, there isn't just one person, especially when considering all of the elements that influence a system.

2.2 Elements that Influence a System

As discussed in Chapter 1 and shown in Figure 1.1, there are many elements that influence a system including: proactive maintenance, operating procedures, technical publications, training programs, equipment design, supply issues, operational tempo and environment, and emergency procedures. The range of issues that directly affect how equipment operates makes it almost impossible for one person to know everything about an asset and what is required for it to meet requirements.

Responsible custodianship means identifying and developing comprehensive failure management strategies.

It doesn't matter what is analyzed—an airplane, nuclear power plant, truck, tank, ship, offshore oil platform, mobile air conditioning unit, tow tractor, jet engine, or a single pump. Whatever the asset is, responsible custodianship means identifying and developing comprehensive failure management strategies.

2.3 Failure Management Strategies

When formulating failure management strategies to maintain assets, organizations typically focus on the development of a proactive maintenance program. However, there are many other failure management strategies that are almost always required to ensure an asset meets requirements. Examples of these are shown in Figure 1.4; they include new operating procedures, updates to technical publications, modifications to training programs, equipment redesigns, supply process changes, enhanced troubleshooting procedures, and updated emergency procedures. Where, then, can the information required be obtained to formulate these solutions?

2.4 Historical Data and the RCM Process

One place to turn is historical data. Historical data is important and can be incredibly useful. But without exception, the kind of data that is generally collected isn't sufficient to answer all the questions in the RCM process—and, thereby, formulate specific solutions. In many cases, the kind of data collected for assets can be likened to baseball statistics. Figure 2.1 presents a season's batting statistics for batter Smith.

Batter	Batting Average (Avg.)	Home Runs (HR)	Runs Batted In (RBI)
Smith	.204	6	21

Figure 2.1 One season's batting statistics for player Smith

Player Smith's batting average is .204, which means the batter gets a hit approximately twice out of every ten at-bats. He has 21 runs batted in (RBI) and six home runs. By reviewing the data, the batting coach can conclude that batter Smith needs improvement. This is valuable information because now the batting coach knows where resources need to be designated—helping to *improve* batter Smith's performance. However, what the batting coach cannot deduce from reviewing the data is what is *causing* batter Smith to perform poorly so the coach cannot formulate *specific* solutions to help the batter improve. For example, should the batter start to swing a little earlier? Or maybe a little later? Or maybe the batter needs to change his stance. The solutions cannot be determined just by evaluating the data.

Historical data for assets is often of the same ilk. For example, a review of bearing data can reveal that 50 bearings were replaced last year—up from 20 last year. From this review, the equipment custodian can conclude that there is a problem regarding the bearing. However, what *specifically caused* the 50 bearing failures cannot be identified simply from reviewing the data. For example, were the bearings greased improperly? Were they not greased at all? Was the wrong grease used? Was there a manufacturing defect? Were the bearings fitted improperly? There are many issues that could specifically cause the bearing failures. So while the data is valuable because it allows an equipment custodian to zero in on problem areas and, thereby, allocate resources where they can be of most benefit, the data doesn't reveal exactly what is causing the bearing problem.

The Use of Historical Data in an RCM Analysis

When historical data is available, it should be employed in the RCM process. For example, historical data is typically very useful for determining items with high failure rates and high maintenance man-hour consumers. The data allows an organization to focus in on problem areas and assists in prioritizing the systems that will be subject to RCM analysis. In this way, resources are allocated where they would be most beneficial.

Where Historical Data Often Falls Short

Historical data can be incomplete because it typically:

- Reports only what failed
- Describes what was done to repair the failure rather than what caused it
- Doesn't describe failures that are currently being prevented or plausible failures that haven't occurred
- Describes failures which may be the effect of some other failure
- Offers inadequate information for determining On-Condition, Restoration, and Replacement task intervals

The use of historical data in an RCM analysis plays a very important role in the application of RCM, but the data is often incomplete and requires further explanation. So, if historical data is often incomplete to perform an RCM analysis, where can an organization turn to get the information?

2.5 Effective Working Groups

None of us is as smart as all of us. Ken Blanchard

Organizations can capture an enormous amount of information by asking the right people; this tool is one of the most valuable tools in any RCM analysis. When a working group is as-

When a working group is assembled, there are typically over 100 years of cumulative experience at an organization's disposal.

sembled, there are typically over 100 years of cumulative experience at an organization's disposal. Because of the vast and varied experience and perspectives represented, the group shares a unique opportunity to formulate solutions that can make a remarkable difference to the organization. By turning to people who know where the improvement opportunities are, skilled facilitators can use RCM principles to consolidate their knowledge and lead experts in formulating solutions that can have a powerful impact on the organization.

The best working group members have significant experience and understand the equipment, operating environment, operational tempo, and equipment requirements.

In order for a working group to be effective, the most knowledgeable and experienced individuals are required. The best working group members have significant experience and understand the equipment, operating environment, operational tempo, and equipment requirements. Suppose an individual is requested to participate in an analysis and management reports it can't afford to have that individual away for a week or two; that is confirmation that the right person has been identified. In fact, the organization can't afford *not* to have the expert in the analysis.

2.6 Benefits of a Facilitated Working Group Approach

More Safe, Cost-Effective, and Technically Defensible Proactive Maintenance

The questions that RCM poses require specific and detailed answers. For example, when trying to determine an On-Condition task to monitor a V-belt for wear, a facilitator may ask a team the following question: *How much time will it take from the point that visual evidence of wear on the V-belt is detectable to the time that the belt breaks?* This span of time is known as the P-F interval and is discussed at length in Chapter 9. The answers to questions like this one are rarely found in historical data because this type of data usually isn't captured and tracked. However, equipment experts–people who work with the equipment every day–are poised to answer the RCM questions most of the time. In this case, a machine operator can usually identify, for example, that it would take six months for the belt to break once cracks and frays are detectable. This is a straightforward example, but many RCM issues can be very complex, which makes it ever more important that experts are allowed to answer the questions. This ensures that the most safe, cost effective, and technically defensible proactive maintenance plan is formulated—that is, the right maintenance is done at the right time.

Results Go Far Beyond Equipment Maintenance

Because they work with the equipment on a day-to-day basis and understand the intricacies of the equipment and the operating environment, working group members understand the vulnerabilities of equipment and the associated processes that lead to equipment failure. This understanding allows them to know where the improvement opportunities are. When asked the right questions, working group members can formulate failure management strategies that go far beyond proactive maintenance (such as changes to operating procedures and updates to technical publications), allowing issues other than proactive maintenance to be addressed.

Working Group Members Learn from Each Other

As stated before, it is almost impossible for one person to know everything there is to know about an asset. Because the cumulative knowledge of a working group is so vast and varied, team members learn from each other during an analysis. Their familiarity, awareness, and understanding of the equipment and the organization grow, allowing their contribution to the organization to become even more valuable. Very often during an RCM analysis, the facilitator gets feedback from a working group member such as "wow, I've been working on this equipment for 20 years and I didn't know that." Even the most seasoned expert learns.

RCM Identifies What an Organization Doesn't Know

Working group members often use experience and judgment to provide answers, but they don't take guesses.

Because the RCM process requires answers to detailed questions, one of its greatest strengths is that it naturally identifies what an organization doesn't know. Gaps in information are documented so that the information can be obtained. Working group members often use experience and judgment to provide answers, but *they don't take guesses.* Facilitators are trained to recognize when a working group doesn't know and appropriate action is taken as a result. For example, an age exploration program may be recommended as a result of RCM analysis or an action item may be issued to obtain further information.

Because the right people were asked the right questions, some of the most successful RCM analyses uncover the issues that have been causing chronic unreliability.

During an analysis, issues requiring additional information are parked until the answers can be found. Just as in life, it is dangerous when you don't know what you don't know. But there is great strength in identifying what you don't know—information can then be obtained while issues are dealt with appropriately in the meantime. Because the right people were asked the right questions, some of the most successful RCM analyses uncover the issues that have been causing chronic unreliability.

Tribal Knowledge is Preserved

In most cases, the knowledge and experience that experts gain over the years isn't formally recorded. So when seasoned experts retire or choose to leave an organization, often the intricacies of their know-how leave with them. Harnessing this information can be incredibly valuable to an organization. The RCM process formally extracts and documents this knowledge so that future generations of equipment experts, and thereby the organization, can benefit from it.

Reduces Human Error

Human error is a widespread problem across the world. History is riddled with fatal disasters caused by it. Equipment is often so complex that there will always be vulnerabilities present that can lead to disasters if not identified and eliminated. On the surface, it may appear that a technician is at fault, but a more detailed inspection may reveal the real cause. For example, if a component is installed backwards, the typical reaction is to blame the technician who installed it. But the backwards installation may actually be an effect of a deeper problem: *the maintenance manual wrongly depicts the position of the component.* In this case, the technician did the job right. The problem is that the technician was tasked with the wrong job.

Consider the following disasters caused by "human error".

NASA's Mars Climate Orbiter Lost on September 23, 1999

The $125 million Mars orbiter was to be a key part in exploring the planet. Two engineering teams working on the project were using different units of measure—English and metric units. As a result, on September 23, 1999, the spacecraft entered a much lower orbit than was intended and the spacecraft was lost. Edward Weiler, NASA's Associate Administrator for Space Science, said in his written statement, "The problem here was not the error. It was the failure of NASA's systems engineering, and the checks and balance in our processes to detect the error."

Helios Airways, Flight 522

Helios Airways, Flight 522 was a Boeing 737-300 that crashed in a hilly area 25 miles north of Athens on August 14, 2005. The aircraft underwent maintenance the night before the accident. The ground crew left a cabin pressurization setting on "manual" mode instead of "auto" mode. As a result, the cabin would not pressurize after takeoff. The crew ignored the cabin altitude warning horn, the passenger oxygen mask deployment indicator, and the master caution switch, and the aircraft continued to climb. The crew then suffered from hypoxia, or oxygen deprivation. The aircraft crashed when it ran out of fuel. All 121 people on board died.

A direct cause cited in the final accident report by the Hellenic Air Accident Investigation and Aviation Safety Board (AAIASB) includes failure of the crew to recognize that the pressurization selector was in manual mode during preflight and performance of the "before start" and "after takeoff" checklists. There were also latent causes cited, including deficiencies in the operator's organization, quality management and safety culture, regulatory authority's inadequate execution of safety oversight, and crew resource management principles not adequately applied.

Since the accident, the AAIASB and the FAA issued several safety corrections. An airworthiness directive was issued requiring a revision of 737 aircraft flight manuals to reflect improved procedures for setting cabin pressurization and crew response to cabin altitude warnings.

Responsible custodians must identify where these kinds of vulnerabilities, also known as latent conditions, exist so steps can be taken to eliminate them. Once again, it is the equipment experts who are in the best position to know where these types of problems are. Equipment experts know what:

- areas of technical manuals that are wrong
- training is inadequate
- checklists are inaccurate
- valves are difficult to operate because of poor lighting
- controls are not accessible
- electrical schematics have errors
- emergency procedures are lacking

These issues, and others like them, can cause failure just as easily as a worn bearing can lead to a bearing failure. Therefore, latent failures must be considered in an RCM analysis.

Accident investigations reconstruct what led to a disaster. They are reactive. Considering potential accident causes in an

RCM analysis proactively focuses on identifying latent failures so that solutions can be formulated to eliminate them. This analysis can make the world a much safer place.

Fosters Relationships Amongst Personnel

It is magical to see what happens when equipment experts from varying backgrounds are assembled together and engage in the RCM process to make decisions about the assets for which they are responsible. It is often the first time, for example, that the operator and the engineer join forces and engage in a formal, structured process to collaboratively formulate solutions about what they have in common—*the asset.* Interaction from varying disciplines fosters a team environment.

Builds Ownership of the Final Results

Because representatives of most disciplines within the organization performed the analysis, the results are much more likely to be embraced. Any changes that occur as a result of RCM are more easily understood. Each area of the organization has an investment in the final results.

Summary

The range of issues that directly affect how equipment operates makes it almost impossible for one person to know everything about an asset, including what is required for the asset to meet requirements. Organizations can capture an enormous amount of information by asking the right people. This tool is one of the most valuable in any RCM analysis. By turning to people who know where the improvement opportunities are,a skilled facilitator can use RCM principles to consolidate a team's knowledge and lead experts in formulating solutions that can have a powerful impact on the organization.

Three

The RCM Operating Context

Answers to the questions posed during an RCM analysis depend on myriad factors such as what the system *is*, how the organization needs it to *perform*, and in what environment it is expected to operate. These and other factors should be clearly defined in the Operating Context before starting an analysis.

This chapter explores:

- What an Operating Context is
- When an Operating Context should be drafted
- What is included in an Operating Context
- The Operating Context as a Living Document

An Operating Context
is a storybook identification of the system to be analyzed.

3.1 What Is an Operating Context?

An Operating Context is a document that includes technical information relevant to the RCM analysis. In essence, it is a storybook identification of the system to be analyzed. Because of the level of detail included, it often acts as a centerlining tool to orient working group members as an analysis begins. The Operating Context also aids in brainstorming. As system details are defined, issues arise that are identified for inclusion in the RCM analysis. The Operating Context also serves as a source of information for validators.

3.2 When Should an Operating Context Be Drafted?

Much of the information recorded in the Operating Context must be sought in various technical documents and may require some research. In the interest of time, the Operating Context is typically drafted by the facilitator before the analysis begins. However, the document is merely a draft. It should be reviewed with the working group before the seven steps of the RCM process are accomplished. During the RCM analysis, the working group can review and edit the Operating Context, as required.

RCM Operating Context

General Information, Purpose and Use

Scope of Analysis

Theory of Operation

Equipment Description

Protective Devices

Operating Context Concerns

RCM Analysis Notes

Future Plans

Working Group Members

Figure 3.1 Sections included in an RCM Operating Context

3.3 What is Included in an Operating Context?

The sections included in an Operating Context are shown in Figure 3.1.

General Information, Purpose and Use

This section includes a description of the equipment to be analyzed. The topics that may be included are:

- A broad description of the organization and the system to be analyzed
- An explanation of the system's role within the organization (e.g., an air dryer system is required to dry compressed air to prevent rusting, pitting, blockages, and freeze-ups of downstream equipment.)
- Number of units in inventory
- How long the system has been in operation or if the equipment is new
- Under what circumstances the system will be operated (e.g., stand-alone, one of four systems runs at one time but is rotated every month)
- Description of the operating environment (e.g., equipment located outside and is exposed to the elements, located in a controlled environment where the temperature varies only between 65–80° Fahrenheit, shipboard use)
- Describe the operational tempo (e.g., 24-hour operation, runs only 6 hours each day)
- How failure of the system affects the organization, if at all (e.g., operations cease and each hour of downtime costs the organization $10,000 in lost production)
- Redundancy (e.g., if the system or any of its components operate in the presence of a backup)

The following example comes from an Operating Context of

a mobile diesel generator. The system subject to RCM analysis is the fuel system.

> The Porter Consolidated Corporation has been commissioned to construct a new power generating facility. The site will house one 300 MW coal-fired power unit. Initial planning estimates that the facility will be operational in three years. Excavation and site preparation are complete.
>
> The Porter Consolidated Corporation has purchased 30 mobile diesel generators to supply power for construction activities. Each single-bearing generator includes control circuitry that provides various indications, warnings, and fault protection. Each generator supplies uninterrupted power to its respective construction area on a 24-hour basis. If a generator fails, construction in the respective area is halted at a cost of $5,000 per hour.
>
> The current operating environment has standard low and high temperatures of –10° F and 100° F, respectively. The generators are maintained by the site equipment maintenance team.

Scope of Analysis

The analysis scope defines the boundaries of the analysis. It details what is included in the analysis. The scope should be clearly defined during the analysis planning stage. The following example is from the mobile diesel generator fuel system.

> The scope of analysis includes the diesel fuel system from the fuel tank assembly to the six fuel injectors. It also includes the fuel level indicator and the low fuel alarm system.

Theory of Operation

The Theory of Operation includes a detailed description of how the system operates, including specific parameters such as operating pressures and temperatures. Parameters should reflect what the organization requires from the asset—*not design specifications.* The following example is from the mobile diesel generator fuel system.

> The mobile diesel generator fuel system stores, cleans, and

delivers fuel to the engine for continuous combustion. The fuel system is required to deliver fuel to the engine at an uninterrupted delivery flow of 2,300–2,700 psi to the engine while operating under load.

Fuel is drawn from a welded steel fuel tank assembly into the primary fuel filter. The primary fuel filter is located downstream of the fuel tank. A differential pressure gauge is mounted on the top of the primary fuel filter and indicates the pressure drop across the filter within +/– 5%. The fuel leaves the primary fuel filter and is drawn into the engine mounted fuel pump. Pressurized fuel then leaves the pump and passes through a secondary fuel filter. The secondary fuel filter is located downstream of the fuel pump.

Once leaving the secondary fuel filter, the pressurized fuel passes through a check valve, which is installed between the secondary filter and the cylinder head, to remove air from the fuel supply line. Under load, pressurized fuel then travels to the injector nozzles which spray diesel fuel into the engine's combustion chamber. Surplus fuel exits at the rear of the head, through a restricted return fitting which maintains system fuel pressure, and returns back to the fuel tank.

All associated fuel lines are 1/4″ rigid lines. Fuel connectors are used to connect the fuel lines to the fuel system components using a cork type gasket. A manually operated fuel shut-off valve (ball valve) is located downstream of the secondary fuel filter to isolate the fuel system and prevent fuel drain-back when the fuel filters are replaced.

An analog fuel level indicator is located on the control panel. Visual reference of fuel quantity is indicated by a needle pointer which moves along the indicated fuel scale graduations. A float assembly located at the top of the fuel tank sends a signal to the fuel level indicator via a sending unit to indicate fuel quantity within +/– 10% of actual. In the event that the fuel level is allowed to drop below 1/8 of a tank, a low fuel alarm sounds.

A fuel filler screen is positioned in the filler neck assembly to prevent debris from entering the fuel tank during servicing. The fuel tank is equipped with a manual drain which allows accumulated moisture and debris to be removed.

Equipment Description

A description of the major components and subassemblies associated with the system should be included in the Operating Context, including component specifications, operational limitations, etc. Major components and subassemblies such as tanks, pumps, motors, filters, heaters, coolers, bypass systems, gauges, and control panel equipment should be described. The following examples are from the mobile diesel generator fuel system.

> Primary and secondary filters: The primary and secondary fuel filters are rated at 30 and 2 microns, respectively. Both fuel filters are equipped with a manual drain, which allows accumulated moisture and debris to be removed without removing the filter. The filters' media is a treated paper element rated for 1,000 hours of operation.

Protective Devices

Protective devices are intended to protect people, the asset, and the organization in the event that something goes wrong. All protective devices associated with the system should be listed and described in the Operating Context. Examples of protective devices are fire suppression systems, high temperature cutoffs, warning lights and alarms, and emergency stops. Below is an example from the mobile diesel generator fuel system.

> Low fuel alarm: In the event that the fuel level is allowed to drop below 1/8 of a tank, a signal is sent via a sending unit to the low fuel alarm, causing the audible alarm to sound.

Operating Context Concerns

Any problems regarding the system should be detailed. These issues may be reverse engineered and included as Failure Modes in the analysis so that recommendations can be formulated. For example, a system may not have detailed troubleshooting procedures or adequate training associated with it. In addition, there may be a component associated with the system that is experiencing chronic failure.

RCM Analysis Notes

Any notes or assumptions that support analysis decisions should be detailed in the *RCM Analysis Notes* section. Examples may include documenting the following issues:

- Failure Effects are written and consequences are assessed assuming that Operators are equipped with a flashlight at all times.
- The oil filter can be fitted with a differential pressure gauge, but it was not purchased with the equipment.
- Operators wear hearing protection at all times so Failure Modes that cause high noise areas do not pose a hearing hazard.
- The term "circuit" used in Failure Modes indicates any part of the electrical circuitry associated with the scope of the Failure Mode such as wiring, switches, and resistors. This excludes protective devices such as circuit breakers and fuses.
- The technical manual referenced during this analysis is TM 1298-65-25, dated 14 September 2009.

Future Plans

Potential or impending changes to the system should be documented (e.g., changes to hardware, training, technical publications, operating procedures, and operational tempo).

Working Group Members

All working members' names, titles, and departments/organizations are documented in the Working Group Members section.

3.4 The Operating Context as a Living Document

During the analysis, the Operating Context is a living document. It is edited as more is learned about the equipment and additional issues come to light.

Chapter 3

Summary

The Operating Context is prepared before the 7 steps of the RCM process are accomplished. It is a brief document and includes various sections that detail the scope of analysis, the system's technical characteristics, any operating context concerns, and assumptions made that influence the working group's decisions. The Operating Context also serves as a source of information for validators.

Four

Functions

What better way to learn about a piece of equipment than to tear it apart piece by piece on paper and identify what its Functions are?

—RCM Working Group Member

Writing Functions is the first step in the RCM process. This chapter explores:

- The importance of writing Functions
- The two types of Functions: Primary and Secondary
- How Functions are classified: Evident or Hidden
- How to compose Evident and Hidden Functions
- Tips regarding Functions

4.1 Why Write Functions?

The first step in the RCM process is to identify Functions. As detailed in Chapter 1, it is vitally important to start with clarity about what performance is required of an asset. Writing

Functions is essential for two reasons:

1. Functions allow an organization to document *specifically* what is required of an asset so that the RCM team can determine if the asset is capable of serving in that manner.

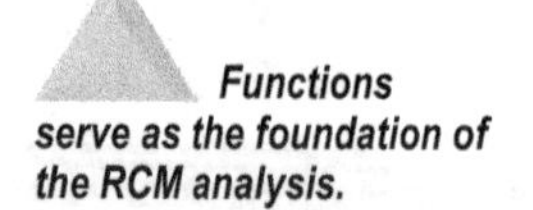

2. Functions serve as the foundation of the RCM analysis. Because Functions document exactly what the organization requires from the asset, solutions can be formulated to ensure assets are maintained to that level.

4.2 Two Types of Functions

As depicted in Figure 4.1, there are two types of Functions: *Primary* and *Secondary.*

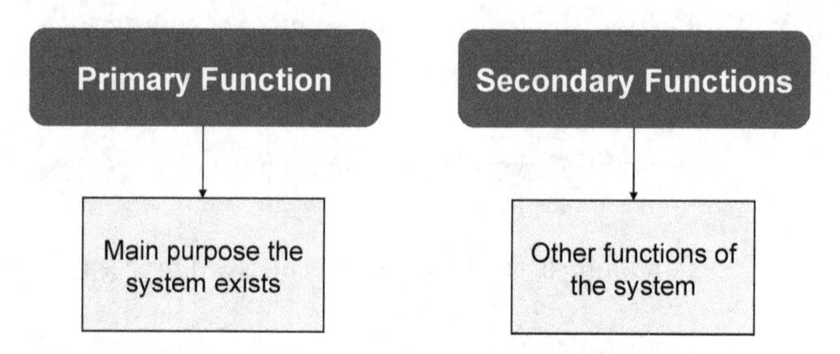

Figure 4.1 Primary and Secondary Functions

The *Primary Function* is the *main purpose that a system exists*. For example, a tow tractor is procured to tow equipment, a boiler is required to produce steam, and a hydraulic system is necessary to provide hydraulic assist. However, tow tractors, boilers, and hydraulic systems are required to perform additional Functions. In the context of RCM, *other Functions of a system* are called *Secondary Functions*. For example, a hydraulic system must maintain the fluid at a specified temperature, filter particulates from the fluid, and indicate the hydraulic pressure.

4.3 Classifying Functions as Evident or Hidden

Functions are classified as either Evident or Hidden.

Evident Functions

Evident Functions are those that, upon failure, become evident to the operating crew under normal conditions. For example, if the battery in an automobile cannot provide adequate electricity to start the car, failure of that Function becomes evident to the operating crew when the car won't start. (In this case, the operating crew is the driver.) Likewise, if the fuel pump fails and cannot deliver fuel to the engine, this failure becomes evident to the operating crew when the car stops running.

Hidden Functions

However, Functions are considered Hidden if, when the Function fails, it does not become evident to the operating crew under normal conditions. In the context of RCM, Hidden Functions are almost always *protective devices*. Protective devices are devices and systems intended to protect people, the asset, and the organization *in the event* that another failure occurs.

For example, a tow tractor includes a low oil pressure safety system that automatically shuts the engine down in the event that the engine oil pressure falls below 10 psi. As described in Figure 4.2, this system is a protective device because it is only needed *in the event* that the oil pressure falls below 10 psi.

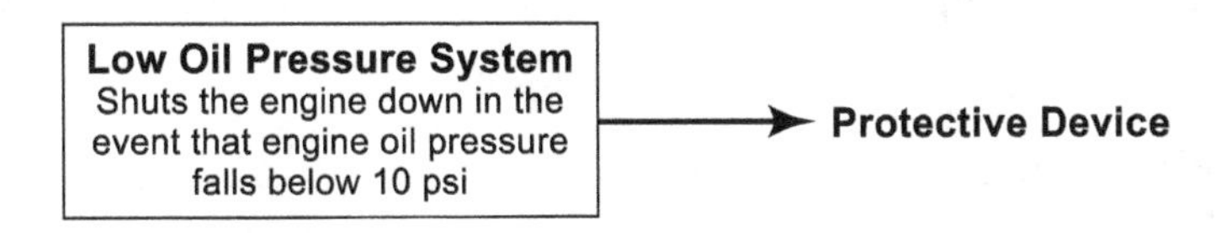

Figure 4.2 Description of a Protective Device

Failure of the low oil pressure safety system is considered a Hidden Function because failure of it *on its own* would not become evident under normal conditions (because normally the oil pressure stays above 10 psi). That is, if the low oil pressure circuit fails, but oil pressure remains above 10 psi, the tow tractor operator wouldn't know it. In the context of RCM, the failure of the low oil pressure system is *hidden*, as depicted in Figure 4.3.

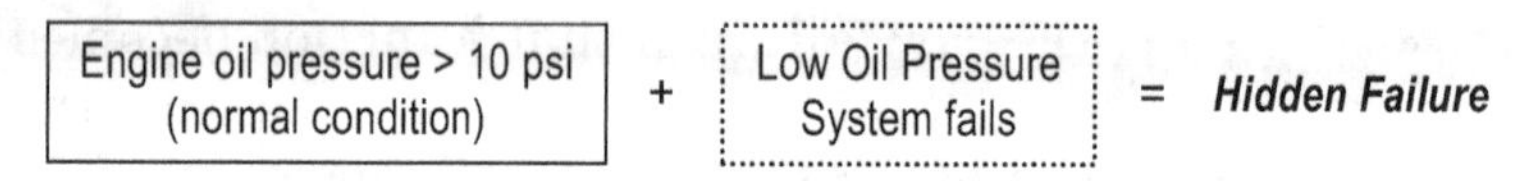

Figure 4.3 Example of a Hidden Failure

The only way the failure of the low oil pressure system becomes evident to the operator is if the oil system pressure drops below 10 psi. In the context of RCM, failure of the low oil pressure system *and* the oil pressure dropping below 10 psi is known as a *multiple failure.*

A multiple failure includes the failure of a protective device *and* another failure. The multiple failure of the low oil pressure system and the oil pressure falling below 10 psi is depicted in Figure 4.4.

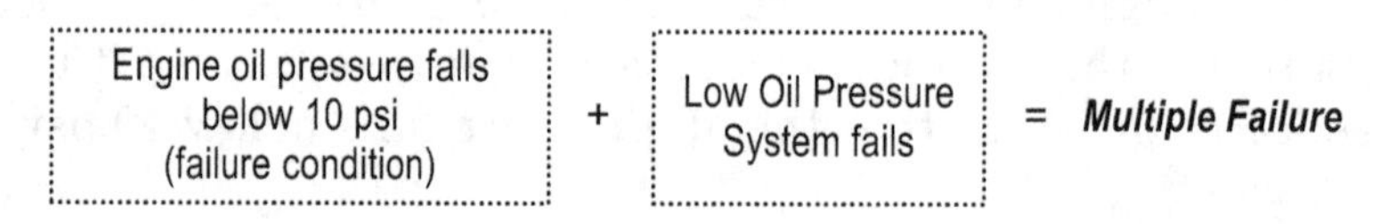

Figure 4.4 Example of a Multiple Failure

When the multiple failure occurs, the engine is not automatically shut down, which allows the engine to run without adequate lubrication. The inadequate lubrication causes the engine to seize.

4.4 Composing Evident and Hidden Functions

When writing Functions, it is important to remember to describe what something is required to do,not what it is.

Required Performance vs. Design Specification Information

When writing Functions, it is important to remember to describe what something *is required to do*, not what it is *designed to do.* At times, working group members require coaching from the facilitator when writing Functions because many working group members typically deal with design aspects of assets rather than the required performance. Consider the following examples that illustrate the difference between required performance and design specification information.

• Design Specification Information: To have a sealed-beam halogen headlight on each side of the front of the tow tractor. This example details what the device *is.*

• RCM Function: To illuminate a maximum of 100 feet in front of the tow tractor during times of darkness and low visibility. This example details what the device *is required to do.*

Composing Evident Functions

Figure 4.5 shows how to write Evident Functions.

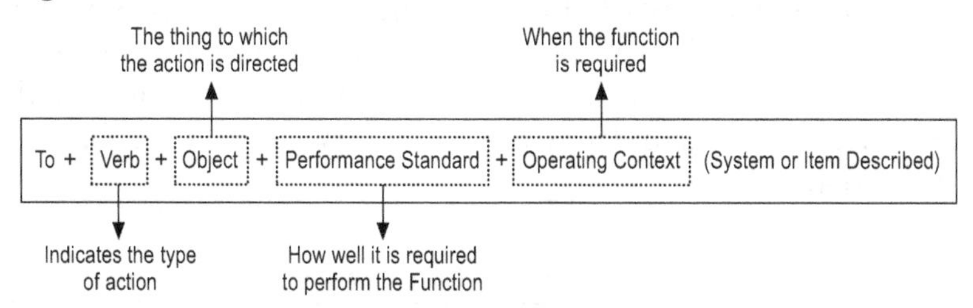

Figure 4.5 How to write Evident Functions

Evident Functions start with the word "To" and incorporate a verb, object, performance standard, and operating context. For Secondary Functions, writing the item described in parentheses at the end is often helpful. When reviewing the list of Functions, the parentheses make it easier to find Functions quickly and to discern which ones have already been written. An Evident Function that follows the model depicted in Figure 4.5 is detailed in the following example:

To display differential pressure across the primary fuel filter within +/- 5% when the system is running. (Fuel filter differential pressure gauge)

Composing Hidden Functions (Protective Devices)

Hidden Functions (protective devices) are written as depicted in Figure 4.6.

To be capable of [...] in the event that [...] (System or Item Described)

Figure 4.6 How to write Hidden Functions (Protective Devices)

A Function for a protective device details what the device needs to be capable of doing and when it is required to operate. A Hidden Function that follows the model depicted in Figure 4.6 is detailed in the following example:

To be capable of interrupting the circuit in the event that the current exceeds 20 amps. (fuse)

4.5 Information Worksheet

Functions are recorded on the Information Worksheet and are designated with a number. The Primary Function is always designated as the first Function. Secondary Functions follow. Functions recorded on the Information Worksheet are depicted in Figure 4.7.

Information Worksheet

	Function		Functional Failure		Failure Mode		Failure Effect
1	Primary Function						
2	Secondary Function						
3	Secondary Function						

Figure 4.7 Functions Recorded on the Information Worksheet

4.6 Primary Functions

Using a tow tractor as an example, consider the following Primary Function (see Figure 4.8a). A Primary Function takes some time for the working group to develop. Additionally, it is common for a Primary Function to include more than one performance standard.

Primary Function of a Tow Tractor	*To tow equipment and objects weighing up to 10,000 pounds at speeds up to 10 mph, on a maximum slope of 2 degrees, as required.*

Figure 4.8a Primary Function

Figure 4.8b dissects the elements of the Primary Function. Notice that there are three performance standards: 10,000 pounds, 10 mph, and 2 degrees. Additionally, the operating context incorporated in the Primary Function is "as required." The term "as required" is often used in Primary and Secondary Functions to describe a Function that isn't constantly required.

Verb	Tow
Object	equipment and objects
Performance Standards	10,000 pounds 10 mph 2 degrees
Operating Context	as required

Figure 4.8b Elements of the Primary Function of the Tow Tractor

4.7 Secondary Functions

Figure 4.9 includes examples of Secondary Functions of a tow tractor.

Secondary Functions of a Tow Tractor	To bring the tow tractor to a controlled stop with no load from a speed of 10 mph over a distance of no more than 18 feet, as required. (hydraulic brakes)
	To maintain engine operating temperature between 140ºF and 220ºF while the engine is running. (cooling system)
	To visually indicate the turning direction of the tractor from the front and rear from a distance of 100 feet, as required. (turn signals)
	To illuminate the cab, as required. (dome light)
	To contain diesel fuel.
	To visually alert personnel of the equipment location during movement or storage at times of darkness or low visibility. (reflective tape)
	To protect designated surfaces from corrosive environment. (paint)
	To identify the unit and inform personnel of unit operating procedures, safety precautions, and pertinent unit data. (placards and stenciling)
	To prevent personnel from slipping when walking or climbing in or on the tow tractor. (non-skid)
	To indicate oil pressure to within +/– 5 psi. (oil pressure indicator)
	To audibly warn of the tractor's presence from a maximum distance of 100 feet, as required. (horn)
	To be capable of shutting down the engine in the event that the engine temperature exceeds 220ºF. (high engine temperature shutdown system)

Figure 4.9 Some Secondary Functions Associated with a Tow Tractor

4.8 General Equipment Features that Typically Warrant Secondary Functions

The technical aspects of assets used from industry to industry vary greatly. Therefore, there is no "checklist" for what to include as Secondary Functions—which demonstrates one of the many reasons why it is essential that a trained facilitator leads the working group. A facilitator is trained to guide the working group so that members can properly identify and write Secondary Functions.

There are, however, general equipment features that typically warrant Secondary Functions. They include:

- Safety features (e.g., reflector tape and non-skid)
- Environmental restrictions (e.g., not to leak more than 10% of total volume of refrigerant in a calendar year)
- Control panels
- Fluid containment (e.g., fuel, oil, and air)
- Monitoring equipment (e.g., gauges, flowmeters, and sight glasses)
- Indicator lights
- Mounting and support structure
- Drains
- Lighting
- Strainers and filters
- Cooling systems
- Access panels and hatches
- Test systems
- Paint
- Aesthetics

- Placards and stenciling
- Startup and shutdown functions
- Remote operation features
- Protective devices (e.g., pressure relief valves, emergency stops, fuses, and high temperature shutdown systems)
- Other elements that affect the health and performance of equipment

4.9 Tips Regarding Functions

The following suggestions may be helpful when writing Functions.

Other Elements that Affect the Health and Performance of Equipment

As discussed in Chapter 1, whatever the equipment, there are myriad issues that can negatively affect equipment performance. For that reason, it is incredibly important that these issues are identified and included in an RCM analysis. Including these issues allows matters to be analyzed using RCM principles so that a technically appropriate and effective solution can be formulated. For some of these issues, it may be necessary to write a Function. Consider the following two examples.

Example 1: Compressor troubleshooting procedures During an RCM analysis of a compressor, the maintenance personnel report that the troubleshooting procedures included in the technical manual are deficient. The lack of detail increases maintenance time when unscheduled maintenance is required to be performed. The working group members choose to write a Function for what they *require* of the troubleshooting procedures so that they can apply the remaining six steps of the RCM process to the Function in order to make recommendations to improve the troubleshooting procedures. The Function they write is: *To efficiently troubleshoot the compressor, as required.*

Example 2: Air start hose During an RCM analysis of a mobile diesel air start unit, the operators report that changing the air start hose assembly cannot be done quickly because it is V-band clamped. They detail that during operations, changing the hose causes delays because the necessary tools must be obtained. They would like to use RCM analysis to formally explore and document a potential solution. As a result, the working group members choose to write a Function for what they require. The Function they write is: *To quickly change the air start hose during operations, as required.*

By including Functions for other elements that affect the health and performance of the equipment, the issues are documented. The remaining six steps of the RCM process can be applied so that an appropriate solution can be formulated. The Validation team members can then consider implementing the solutions.

Containment Functions and Performance Standards

When writing containment Functions, record performance standards. For example, if there is an acceptable amount of leakage, include it in the Function, as detailed in the following example:

To leak less than 20 percent of produced air.

When no amount of leakage is acceptable to the organization, *an absolute containment Function* is written. Absolute containment Functions do not include a performance standard which implies that no leakage is acceptable. An example of an absolute containment function is:

To contain fuel.

Gauges and other Monitoring Equipment

Gauges and other monitoring equipment are included on equipment to monitor various parameters. When writing Functions for gauges and other monitoring equipment, performance standards should reflect how precise the organization requires the information to be. The following Function is for a boiler steam pressure gauge. Note that the performance standard is +/–5% of actual pressure.

To indicate the boiler steam pressure within +/–5% of actual pressure. (Boiler steam pressure gauge)

Startup and Shutdown Functions

The Primary Function is typically written assuming that the asset is up and running, as depicted in Figure 4.10.

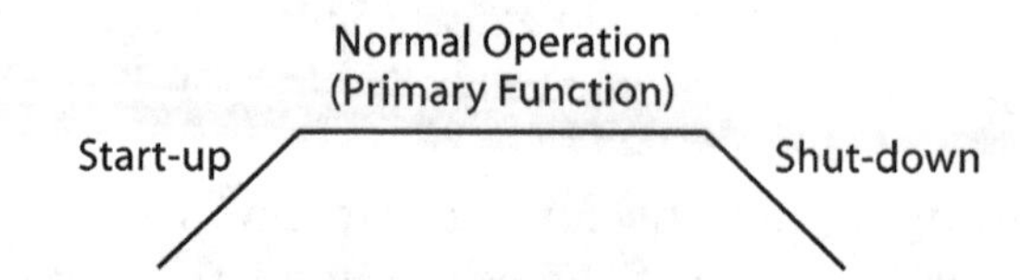

Figure 4.10 Startup and Shutdown features as Secondary Functions

When there are startup and shutdown features, they should be recorded on the Information Worksheet as Secondary Functions, as depicted in Figure 4.11 as Functions 2 and 3.

Information Worksheet

	Function	Functional Failure	Failure Mode	Failure Effect
1	To tow equipment and objects weighing up to 10,000 pounds at speeds up to 10 mph, on a maximum slope of 2 degrees, as required.			
2	To start up the tow tractor, as required.			
3	To shut down the tow tractor, as required.			

Figure 4.11 Startup and Shutdown as Secondary Functions included on the Information Worksheet

Quantifying Performance Standards

Sometimes it is very difficult to quantify performance standards. For example, consider the following Function of an oil reservoir sight glass:

To indicate the compressor oil level.
(Compressor oil reservoir sight glass)

It is very difficult to quantify how well the sight glass is required to indicate. In this case, and others like it, it is acceptable not to include a performance standard at all.

How to Decide if a Function Should be Written

Hundreds, sometimes thousands, of components are associated with an asset that is subject to RCM analysis. Is a Function written for all components?

It is important to remember that RCM is a Function-based process. That is, the working group focuses on what the equipment is *required to do*, not on what it *is*. Because most components contribute to higher level Functions associated with the asset, Functions are not written for all components. The flowchart depicted in Figure 4.12 can be used to determine if a Function should be written.

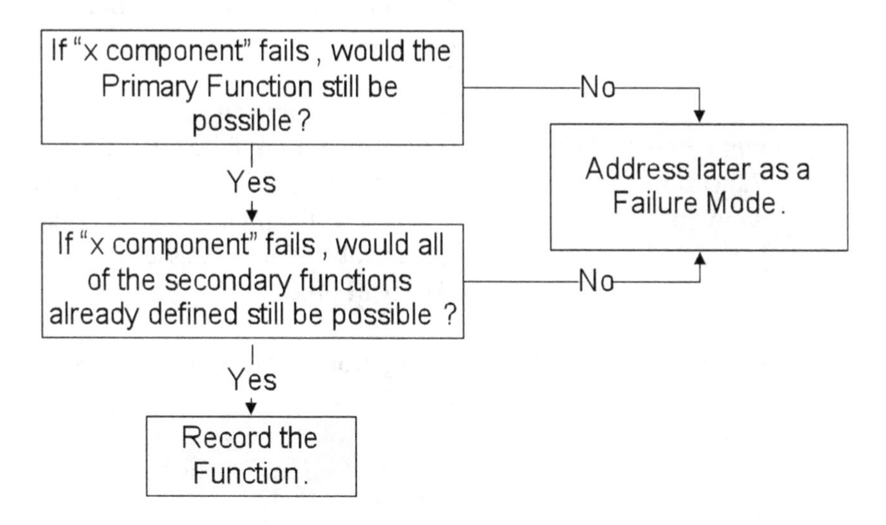

Figure 4.12 Flowchart Used to Determine if a Function Should be Written

Information Worksheet

	Function	Functional Failure	Failure Mode	Failure Effect
1	To tow equipment and objects weighing up to 10,000 pounds at speeds up to 10 mph, on a maximum slope of 2 degrees, as required.			
2	To contain diesel fuel.			
3	To indicate oil pressure to within +/- 5 psi.			

Figure 4.13 Information Worksheet for How to Write a Function Example

Function Exercise

Using the flowchart in Figure 4.12 and based upon the Information Worksheet depicted in Figure 4.13, determine if Functions for the pintle hook and the fuel gauge should be included in the Information Worksheet.

> Pintle Hook: Should the working group write a Function for the pintle hook? No, because if the pintle hook failed, the Primary Function would no longer be possible. That is, the ability to tow would be lost. This doesn't mean that the pintle hook isn't subject to RCM analysis. The pintle hook will be addressed as a Failure Mode during Step 3 of the RCM process.
>
> Fuel Gauge: Should the working group write a Function for the fuel gauge? Yes, because if the fuel gauge failed, towing would still be possible. Additionally, the other Secondary Functions would also still be possible. Therefore, a Function of the fuel gauge should be recorded in the RCM analysis.

Summary

Writing Functions is the first step in the RCM process because it is vitally important to start with clarity about what performance is required of an asset. Functions allow an organization to document *specifically* what is required of an asset so that the RCM team can determine if the asset is capable of serving in that manner. Additionally, Functions serve as the foundation of the RCM analysis. Because Functions document exactly what the organization requires from the asset, solutions can be formulated to ensure assets are maintained to that level.

There are two types of Functions: *Primary* and *Secondary*. The Primary Function is the main purpose the system exists. Secondary Functions include other Functions of the system. Functions are classified as either Evident or Hidden. Evident Functions are those that, upon failure, become evident to the operating crew under normal conditions. A Hidden Function is one that, when it fails, does not become evident to the operating crew under normal conditions. Functions are recorded on the Information Worksheet and are designated with a number. The Primary Function is always designated as the first Function. Secondary Functions follow.

Five

Functional Failures

Writing Functional Failures is the second step in the RCM process. This chapter explores:

- What a Functional Failure is
- The two types of Functional Failures: Total and Partial
- How to compose Functional Failures

5.1 What Is a Functional Failure?

Nowlan and Heap define a Functional Failure as an *unsatisfactory condition.* A Functional Failure is the inability to fulfill a Function.

Functional Failure

The inability to fulfill a Function.

5.2 Two Types of Functional Failures

As depicted in Figure 5.1, there are two types of Functional Failures: *Total and Partial.*

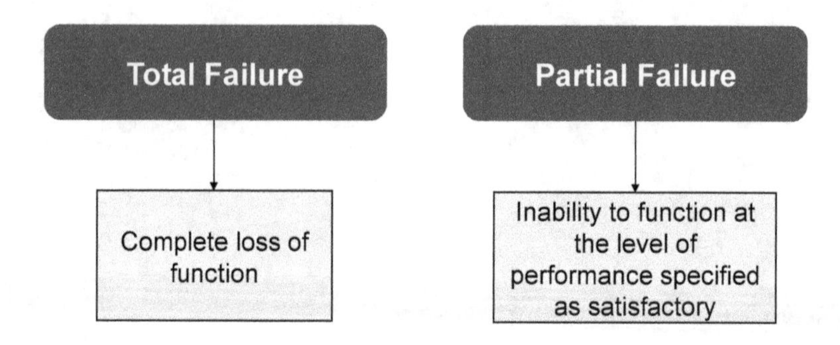

Figure 5.1 Total and Partial Failures

A *Total Failure* means the Function is completely unable to be performed. *A Partial Failure* describes how the Function is still possible but performance is at an unsatisfactory level.

5.3 Composing Functional Failures

Composing Functional Failures is the most straight-forward step in the RCM process. Functional Failures are written for each Function. Each Function is then reviewed to identify the unsatisfactory condition (or conditions) that can arise.

Information Worksheet

	Function		Functional Failure		Failure Mode	Failure Effect
1	Primary Function	A	Total Functional Failure			
		B	Partial Functional Failure			
		C	Partial Functional Failure			
2	Secondary Function	A	Total Functional Failure			
		B	Partial Functional Failure			
3	Secondary Function	A	Total Functional Failure			

Figure 5.2 Total and Partial Failures are documented on the Information Worksheet

Functional Failures

Functional Failures are recorded on the Information Worksheet and are designated by a letter, as depicted in Figure 5.2.

There can be more than one Functional Failure for each Function. Typically the Primary Function includes more than one Functional Failure. Secondary Functions can also have more than one Functional Failure. Some Secondary Functions only include one Functional Failure in which case the Functional Failure is usually just the negative of the Function. When referencing a Functional Failure, the two-character Functional Failure identifier is used. For example, in Figure 5.2, the first Functional Failure associated with Function 1 is 1A. The next Functional Failure is 1B, and so on.

Figure 5.3 illustrates examples from the fuel system of a diesel generator.

The following discussion explains the Functional Failures for the fuel system of the mobile diesel engine generator detailed in Figure 5.3.

Function 1 (Primary Function)

The Primary Function includes two Functional Failures. The first Functional Failure (1A) is total failure. That is, the unit is completely unable to deliver diesel fuel. In the second Functional Failure (1B), diesel fuel can still be delivered but at a delivery pressure of less than 2,300 psi. These are examples of Total and Partial Failure, respectively.

Function 2

As discussed in Chapter 4, Function 2, *To contain diesel fuel*, is considered an absolute containment Function which means any amount of a leak is considered failure by the organization. Therefore, there is only one Functional Failure (2A) associated with this Function, which is total failure.

Function 3

Function 3 is another example of a Function that can experience total and partial failure. Functional Failure 3A details total failure of the fuel filter differential pressure gauge—it is completely unable to display differential fuel pressure. On the other hand, Functional Failures 3B and 3C describe how the gauge

Information Worksheet

	Function		Functional Failure		Failure Mode		Failure Effect
1	To deliver diesel fuel to the engine at an uninterrupted delivery pressure flow of 2,300-2,700 psi while operating under load	A	Unable to deliver diesel fuel				
		B	Delivers diesel fuel to the engine at a delivery pressure of less than 2,300 psi while operating under load				
2	To contain diesel fuel	A	Unable to contain diesel fuel				
3	To display differential fuel pressure across the primary fuel filter within +/- 5%. (Fuel filter differential pressure gauge)	A	Unable to display differential fuel pressure across the primary fuel filter				
		B	Displays actual differential fuel pressure across the primary fuel filter within more than +5%				
		C	Displays actual differential fuel pressure across the primary fuel filter within more than -5%				
4	To isolate the fuel system, as required. (Manual fuel shutoff valve)	A	Unable to isolate the fuel system, as required.				
5	To be capable of sounding an audible alarm in the event that the fuel level drops below 1/8 of a tank. (Low Fuel Alarm System)	A	Incapable of sounding an audible alarm in the event that the fuel level drops below 1/8 of a tank				
		B	Falsely sounds the low level fuel alarm system				

Figure 5.3 Examples of Total and Partial Failures documented on the Information Worksheet

provides a reading, but the reading is outside the tolerance specified by the organization as acceptable. Therefore, Functional Failures 3B and 3C are considered partial failures.

Function 4

Function 4 warrants just one Functional Failure 4A. The working group identified that only total failure is plausible.

Function 5

Function 5 is a Function of a protective device. Functional Failure 5A details the total Functional Failure. In this case, the system is completely incapable of sounding an audible alarm in the event that the fuel level drops below 1/8 of a tank. However, Functional Failure 5B illustrates how the protective device can fail in the opposite way—it can falsely sound the low level fuel alarm system.

Summary

Writing Functional Failures is the most straightforward step in the RCM process. Functional Failures are written for each Function. Functions may have just one Functional Failure or may include several. Functional Failures lead to the next step in the RCM process—identifying Failure Modes.

Failure Modes

Identifying Failure Modes is Step 3 in the RCM process. This chapter explores:

- What a Failure Mode is
- Documenting Failure Modes and the Information Worksheet
- How to compose Failure Modes
- What Failure Modes should be included in an RCM analysis
- How detailed Failure Modes should be written
- Identifying Failure Modes for each Functional Failure
- Tips for writing Failure Modes

6.1 What Is a Failure Mode?

A Failure Mode is *what specifically causes a Functional Failure.* Failure Mode identification allows the working group to determine what causes *conditions that are unsatisfactory.* In other words, the working group identifies Failure Modes that could *cause* the Functional Failures listed in the Information Worksheet.

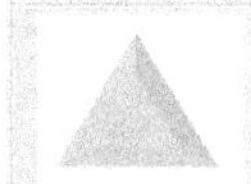

Failure Mode

What specifically causes a Functional Failure.

Typical Failure Modes address issues such as:

- Normal wear
- Corrosion
- Erosion
- Fatigue
- Abrasion
- Lubrication dissipates
- Lubrication deteriorates
- Misalignment
- Dirt build-up
- Drifts out of adjustment
- Leaks

However, RCM isn't just about equipment maintenance. As discussed in Section 1.2 of Chapter 1, there are other elements that affect the health and performance of equipment, such as operating procedures, technical publications, and equipment design. If any of these elements are inadequate or ill conceived, they can adversely affect equipment performance just as easily as a corroded component can. These issues may include:

- Deficiencies in technical manuals
- Incomplete operating procedures
- Proper tools not available
- Inadequate equipment design
- Inaccurate checklists

Working group members are in the best position to identify these shortcomings. Including issues in an RCM analysis, such as mistakes in technical manuals, allows solutions to be formulated.

Likewise, as discussed in Section 2.6 of Chapter 2, human error is a serious issue in the field of asset management and is a widespread problem across the world. On the surface, it may appear that a technician is at fault. However, further examination may reveal vulnerabilities, also known as latent conditions, in the system. For example, an inappropriate operating procedure could lead an operator to make an error. Responsible custodians must identify latent conditions so steps can be taken to eliminate them. Including latent conditions in an RCM analysis as Failure Modes allows solutions to be formulated. Latent conditions may include:

- An incorrect electrical schematic
- The absence of an emergency procedure
- A confusing control panel design

6.2 Failure Modes and the Information Worksheet

The working group identifies what Failure Modes could cause each Functional Failure.

In Step 3 of the RCM analysis, the working group identifies what Failure Modes could cause each Functional Failure. These Failure Modes are then recorded on the Information Worksheet and are designated by a number, as depicted in Figure 6.1. There can be any number of Failure Modes associated with a Functional Failure. Typically, the Functional Failures associated with the Primary Function have the most Failure Modes.

When referencing a Failure Mode, the three-character Failure Mode identifier is used. For example, in Figure 6.1, the first Failure Mode associated with the Functional Failure 1A is 1A1.

Information Worksheet

	Function		Functional Failure		Failure Mode		Failure Effect
1	Primary Function	A	Total Functional Failure	1	Failure Mode		
		A	Partial Functional Failure	2	Failure Mode		
		A	Partial Functional Failure	3	Failure Mode		
		B	Partial Functional Failure	1	Failure Mode		
		C	Partial Functional Failure	1	Failure Mode		
2	Secondary Function	A	Total Functional Failure	1	Failure Mode		
		B	Partial Functional Failure	1	Failure Mode		
3	Secondary Function	A	Total Functional Failure	1	Failure Mode		

Figure 6.1 Failure Modes documented on the Information Worksheet

The next Failure Mode is 1A2. The first Failure Mode associated with Functional Failure 2A is 2A1, and so on.

6.3 Composing Failure Modes

As depicted in Figure 6.2, Failure Modes consist of a *noun*, a *verb*, and when necessary, operating context.

Noun + **Verb** + [as necessary: **operating context**]

Figure 6.2 How to write Failure Modes

The following Failure Modes include a *noun* and a *verb*.

- Compressor disc fatigues.
- Power turbine blade fatigues.

At times, Failure Modes must be clarified with operating context to ensure that an appropriate failure management strategy can be formulated, especially for items with more than one Failure Mode. Consider the following sets of Failure Modes depicted in Figure 6.3. The operating context is represented in italic text.

In the first set of Failure Modes, without the operating context included, the Failure Mode would be the same: *hydraulic line chafes*. This concept is true for the oil filters and foreign objects in an engine air inlet.

Figure 6.4 shows how vastly different the failure management strategies can be for the same item. For example, if a hydraulic line chafes due to normal equipment vibration, RCM analysis may determine to visually inspect the hydraulic line every 25 hours of operation. However, if a hydraulic line chafes due to improper routing, then RCM analysis may determine to augment the training program so that hydraulic lines are routed properly. Similarly, different failure management strategies are documented for the oil filters and foreign objects intruding the engine air inlet.

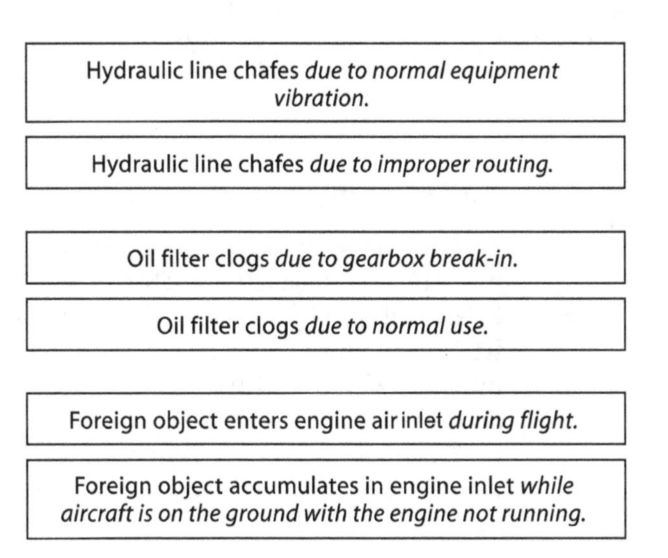

Figure 6.3 Three sets of Failure Modes with operating context in italics

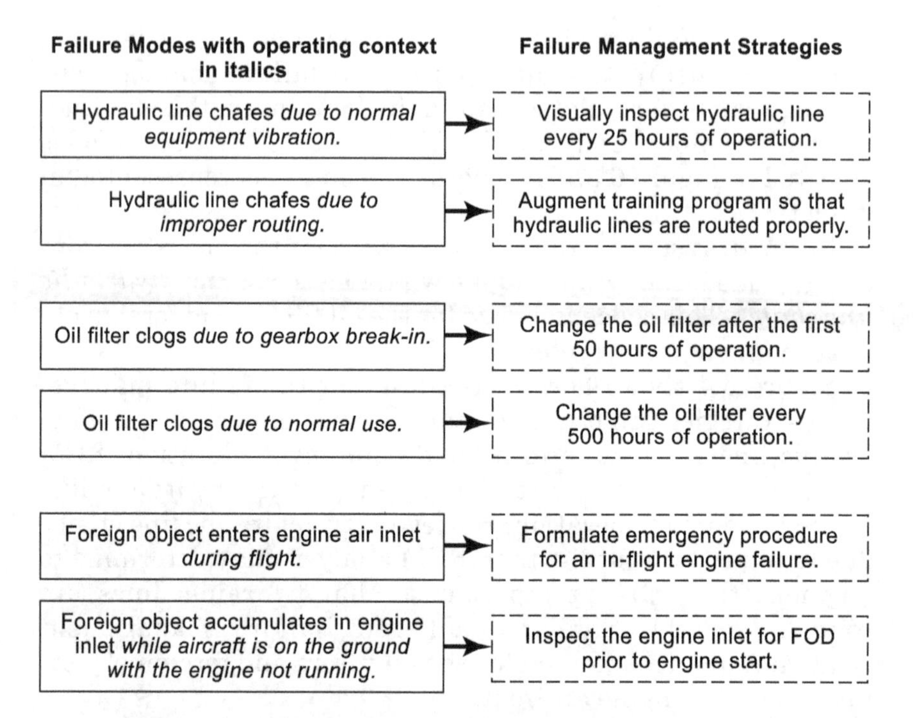

Figure 6.4 Three sets of Failure Modes that include operating context and the Failure Modes' associated Failure Management Strategies

RCM provides specific guidelines for determining what Failure Modes should be included in an RCM analysis.

6.4 What Failure Modes Should Be Included in an RCM Analysis?

It is often wrongly believed that *all* Failure Modes associated with the system being analyzed must be recorded. On the contrary, RCM provides specific guidelines for determining what Failure Modes should be included in an analysis. The flowchart depicted in Figure 6.5 can be used to determine if a Failure Mode should be included in an RCM analysis.

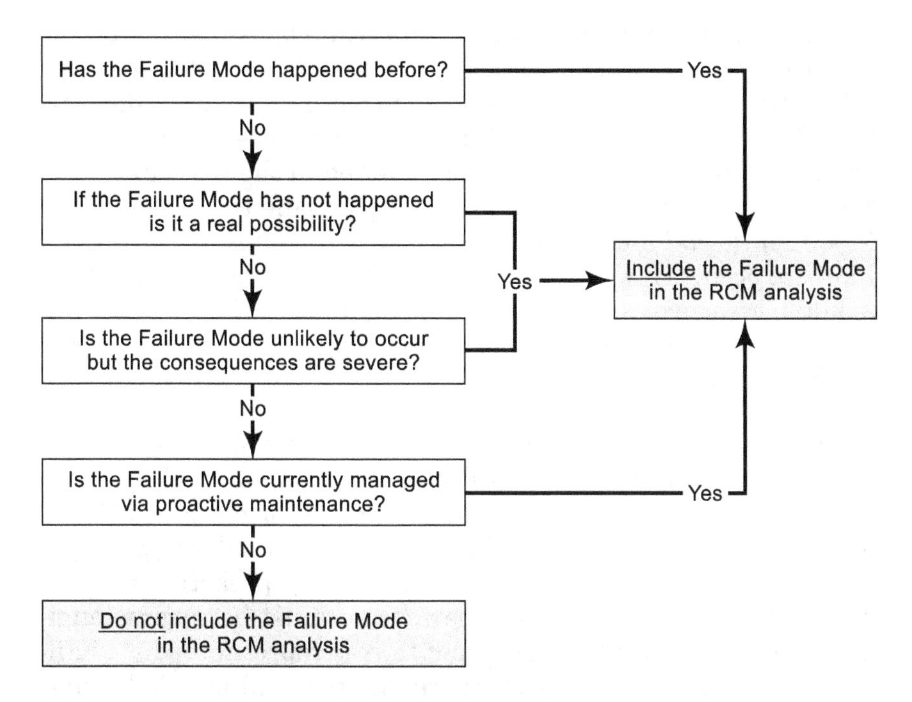

Figure 6.5 Flowchart to determine what Failure Modes to include in an RCM analysis

- Has the Failure Mode Happened Before?

If a Failure Mode has happened before, it should be included in the RCM analysis.

- If the Failure Mode has Not Happened, Is it a Real Possibility?

Oftentimes, some plausible Failure Modes haven't occurred yet. They may have not yet occurred because the Failure Modes are currently being predicted or prevented. In any case, these Failure Modes should be included in the RCM analysis.

- Is the Failure Mode Unlikely to Occur But the Consequences are Severe?

At times, a Failure Mode is brought up during a working group session, but the consensus is that the Failure Mode is unlikely to occur. When this happens, the consequences of the Failure Mode must be considered. If, for example, someone could be injured or killed, an environmental law could be breached, or serious operational consequences could occur, then the Failure Mode should be included in the RCM analysis. For example, a working group may determine that the Failure Mode *Main fuel manifold nozzles clog due to normal use* is unlikely to occur. However, they determine that the Failure Mode would have serious operational consequences. Therefore, based upon Failure Mode guidelines, they choose to include it in the RCM analysis.

- **Is the Failure Mode Currently Managed via Proactive Maintenance?**

This is an important Failure Mode guideline. First, the working group has the opportunity to brainstorm Failure Modes and populate all Functional Failures. Then, in order to preserve the integrity of the zero-based analysis, any proactive maintenance that is currently being performed must be reverse-engineered to ensure that each current task is represented by a Failure Mode. Reverse-engineering means the Failure Mode, that the current task is intended to manage, is identified. Examples of reverse engineering are depicted in Figure 6.6.

Reverse engineering current proactive maintenance tasks is important because it provides an opportunity to ensure Failure Modes haven't been missed and it allows the current proactive maintenance program to be sanity checked. For the current proactive maintenance program, RCM analysis may:

1. Verify the task and the associated interval
2. Verify the task but change the associated interval
3. Recommend an entirely different proactive maintenance task
4. Recommend a different failure management strategy altogether

Examples of the four changes RCM may recommend as a result of reverse engineering current maintenance tasks are depicted in Figure 6.7.

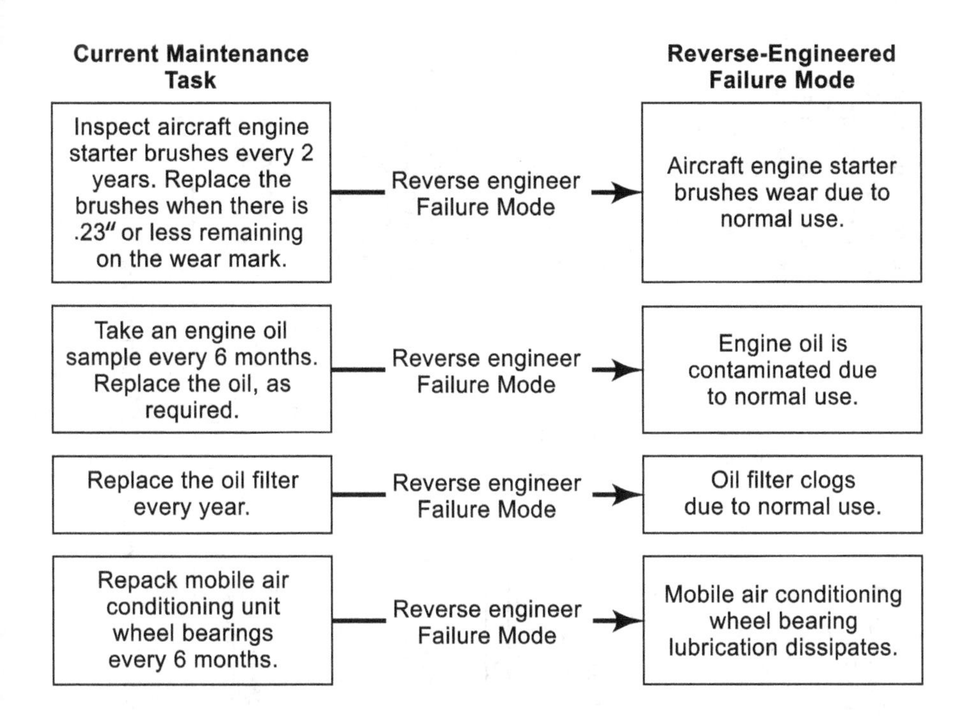

Figure 6.6 Current Maintenance Tasks are Reverse-Engineered to the Failure Modes the tasks are intended to manage

In the first example regarding the engine starter brushes, the result of the RCM analysis is that the current maintenance task remains the same. In the next example, the engine oil sampling interval is shortened from six months to every three months. For the oil filter, instead of changing the oil filter on a fixed interval, the working group recommends monitoring the differential pressure and only changing the filter when the differential pressure exceeds nine psig. Finally, the task for repacking the mobile air conditioning unit's wheel bearings every six months is eliminated. RCM analysis ensures that the proactive maintenance tasks most technically appropriate and worth doing are identified. The concepts of *technically appropriate* and *worth doing* are discussed in detail in Chapter 9.

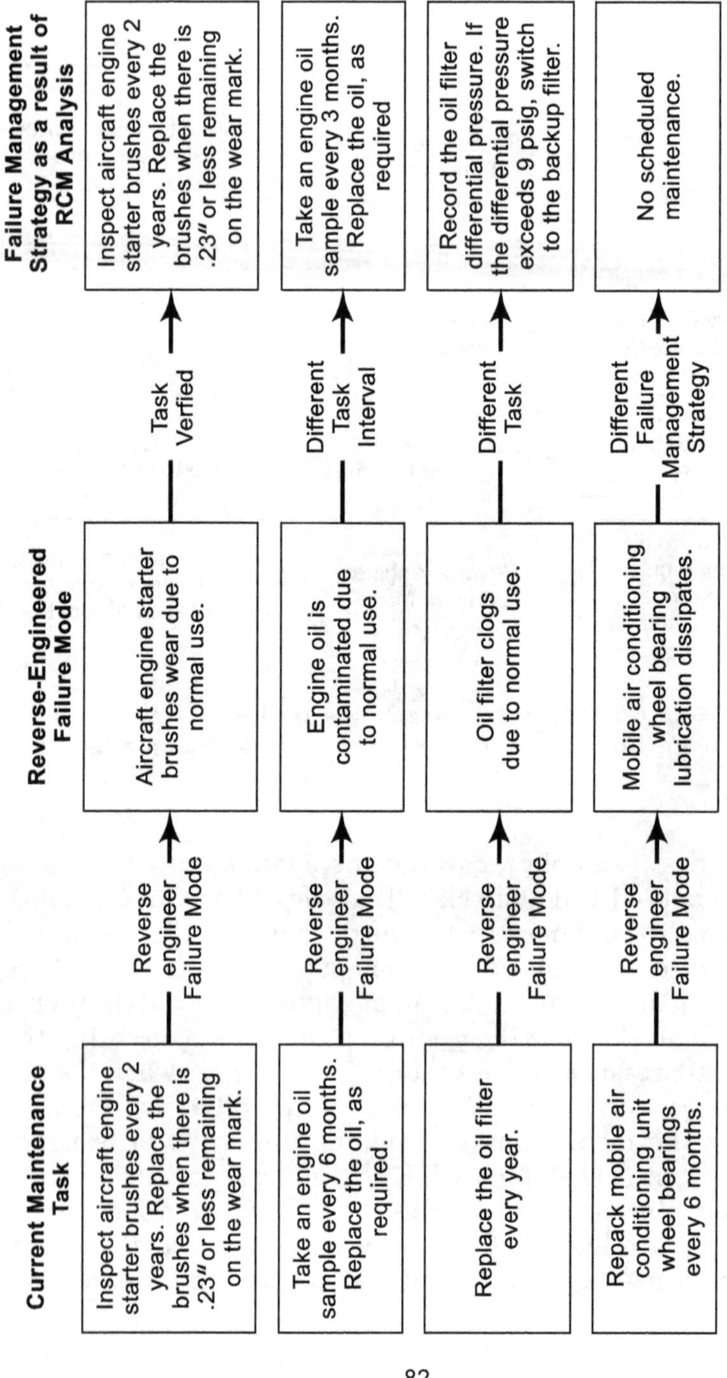

Figure 6.7 Examples of recommended changes as a result of reverse engineering current maintenance tasks

6.5 How Detailed Should Failure Modes be Written?

Failure Modes should be written in enough detail so that an appropriate failure management strategy can be developed.

Failure Modes should be written in enough detail so that an appropriate failure management strategy can be developed. The following discussion explores this concept.

Lower Level Failure Modes

Consider a mobile power unit powered by a diesel engine, with the Failure Mode *Mobile power unit engine fails*. Is this Failure Mode written in enough detail for a solution to be formulated?

In the context of RCM, the Failure Mode *Mobile power unit engine fails* is written at too high a level. It doesn't have enough detail to formulate a failure management strategy. Figure 6.8 details how the Failure Mode can be broken down to more detailed Failure Modes (in the context of RCM, a lower level) so that a working group can make decisions on how to manage the Failure Modes. This example demonstrates another reason why it is important to work with a trained facilitator who is prepared to help working group members get to the right level of detail when writing Failure Modes.

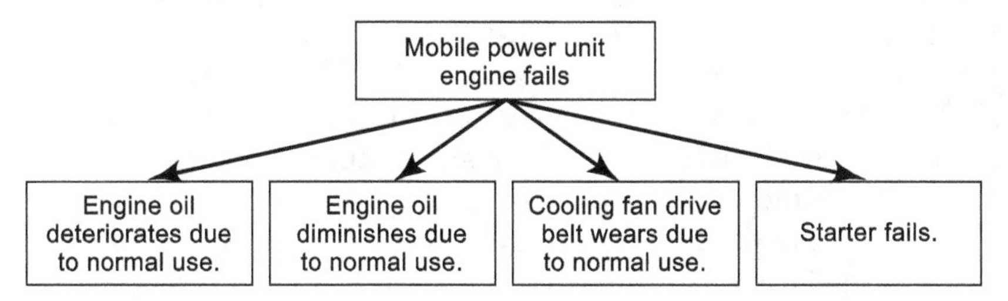

Figure 6.8 Engine Failure Mode broken down to lower level Failure Modes so appropriate Failure Management Strategies can be developed

Figure 6.9 details the failure management strategies identified for the lower-level Failure Modes regarding the mobile power unit engine. The detail included in each Failure Mode allows four distinct failure management strategies to be formulated: changing the engine oil every six months, checking the engine oil level prior to use, visually inspecting the cooler fan drive belt every 500 operating hours, and no scheduled maintenance.

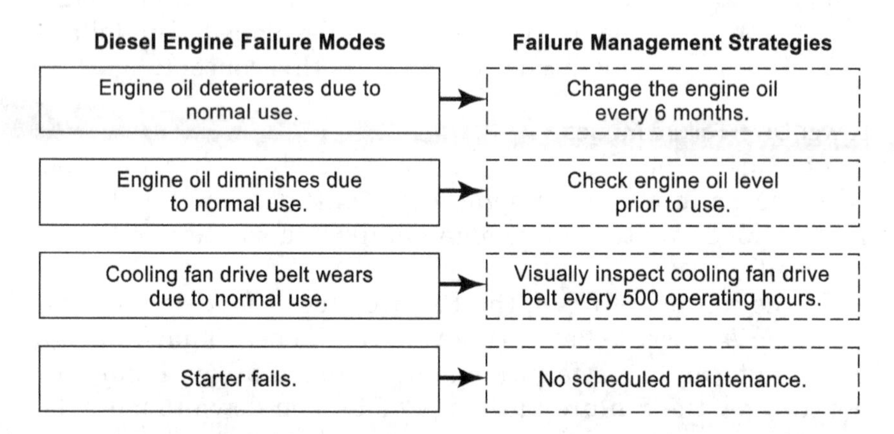

Figure 6.9 Engine Failure Modes and their associated failure management strategies

Higher Level Failure Modes

At times, it *is* appropriate to write a Failure Mode at a higher level when less detail is required to formulate an appropriate failure management strategy. In many cases, the word *fails* is used in higher-level Failure Modes, implying that one Failure Mode is written to account for several Failure Modes associated with a component or subsystem. Consider the following examples:

> **Example 1: Starter motor** In most cases, a vehicle owner would not do any proactive maintenance on a starter motor. If detailed Failure Modes were written for the starter motor, the Failure Modes would include items such as electrical connections and the armature. However, since it is intuitively acknowledged that there is no proactive task to be performed, a high-level Failure Mode can be written as *Starter motor fails.*

Example 2: Protective Device—Hazard Flashers

When writing Failure Modes for protective devices, a high-level Failure Mode is typically written. Consider the hazard flashers on a truck. There are myriad Failure Modes that could occur within the circuit of the hazard flashers, such as the switch, wiring, and lamps. However, the likely failure management strategy for the flashers is to check them from time to time. In this case, it is appropriate to write a high-level Failure Mode as follows: *Hazard flashers circuit fails.*

Considering a Failure Mode as a *road* and a failure management strategy as the *destination*

Working group members need to be on the right track to get to a solution. Therefore, working group members must identify Failure Modes at the right level in order to identify an appropriate failure management strategy.

Categories of Failure Modes to Consider During an RCM Analysis

As already mentioned, RCM isn't just about maintenance. RCM can be used to derive a host of other solutions. Obviously equipment-related Failure Modes must be considered. But in addition, Failure Modes as a result of operating environment issues, sustaining elements, and human error must also be considered. Considering myriad issues during an RCM analysis is one of the elements that make RCM so powerful. Below are sample Failure Modes for each of the four Failure Mode categories: equipment related, operating context issues, sustaining elements, and human error.

Equipment related

- Any 1 of 4 interstage pressure transmitters fails open.
- Battery charge depleted due to normal use.
- Start circuit fails open.
- Paint deteriorates due to normal use.
- Pressure relief valve sticks closed.

- Sediments collect and settle in the bottom of the mud drum due to normal use.
- Electric motor fails.
- Blower sheave wears due to normal use.
- Hydraulic line support loosens due to normal vibration.

Operating Environment Issues

- Moisture develops in bleed air servo filter and freezes in freezing temperatures.
- Ambient temperature varies significantly such that accumulator pressure decreases below required pressure
- Rodents accumulate in exhaust system.

Sustaining Elements

- There is a lack of sufficient battery chargers.
- Steam drum access panel bolts are overtorqued due to lack of torque sequence in Technical Manual AR-32804
- Wrench particular to the backflush valve is unavailable.
- Work package FR54 does not adequately outline the repair steps for a fuel cap grounding strap.

Human Error

- Battery electrolyte is overfilled during maintenance.
- V-band clamp on aircraft end of air start hose damaged by clamp frequently being dropped on the ground.
- Stator temperature alarm setting is set too high.
- Foreign object is left in driveshaft area.
- Overrated fuse installed.

- Aircraft inadvertently damaged by support equipment.
- Fuel line is improperly routed.
- Hydraulic fill cap not replaced properly after servicing.

6.6 Identifying Failure Modes for Each Functional Failure

In Step three of the RCM process, the facilitator guides the working group to identify Failure Modes that cause each Functional Failure. The Failure Modes are documented on the Information Worksheet. Figure 6.10 identifies Failure Modes for each Functional Failure identified for the fuel system of a mobile diesel generator.

When facilitating the RCM analysis, the facilitator asks the working group: *What specifically causes each Functional Failure?* For example, for Functional Failure 1A, the facilitator asks the working group: *What specifically causes the fuel system to be unable to deliver fuel?* One answer is that the *diesel fuel pump fails*, which is documented as Failure Mode 1A1 in the example set forth in Figure 6.10. Failure Modes for the remaining Functional Failures are included in the example.

6.7 Tips for Writing Failure Modes

The following notes may be helpful when writing Failure Modes.

Don't Write Two Failure Modes in One.

For example, it is incorrect to write the following Failure Mode because it contains two Failure Modes in one:

Fuel filter clogs or paper element deteriorates.

Instead it should be split up into two different Failure Modes:

Fuel filter clogs
Fuel filter paper element deteriorates.

Information Worksheet

	Function		Functional Failure		Failure Mode	Failure Effect
1	To deliver diesel fuel to the engine at an uninterrupted delivery pressure flow of 2,300-2,700 psi while operating under load.	A	Unable to deliver diesel fuel.	1	Diesel fuel pump fails.	
				2	Diesel generator is serviced with incorrect fuel.	
		B	Delivers diesel fuel to the engine at a delivery pressure of less than 2,300 psi while operating under load.	1	Primary fuel filter clogs due to normal use.	
				2	Fuel injectors clog due to normal use.	
				3	Primary fuel filter media deteriorates due to normal use.	
				4	Water and debris accumulate in the bottom of the primary or secondary fuel filter due to normal use.	
2	To contain diesel fuel.	A	Unable to contain diesel fuel.	1	Any fuel line connection loosens due to vibration.	
3	To display differential fuel pressure across the primary fuel filter within +/- 5%. (Fuel filter differential pressure gauge)	A	Unable to display differential fuel pressure across the primary fuel filter.	1	Fuel filter differential pressure gauge fails.	
		B	Displays actual differential fuel pressure across the primary fuel filter within more than +5%.	1	Fuel filter differential pressure gauge drifts out of adjustment high.	
		C	Displays actual differential fuel pressure across the primary fuel filter within more than -5%.	1	Fuel filter differential pressure gauge drifts out of adjustment low.	
4	To be capable of sounding an audible alarm in the event that the fuel level drops below 1/8 of a tank. (Low Fuel Alarm System)	A	Incapable of sounding an audible alarm in the event that the fuel level drops below 1/8 of a tank.	1	Low fuel alarm system fails open.	
		B	Falsely sounds the low level fuel alarm system.	1	Low fuel alarm system fails closed.	

Figure 6.10 Failure Modes identified for each Functional Failure for the fuel system of a mobile diesel generator

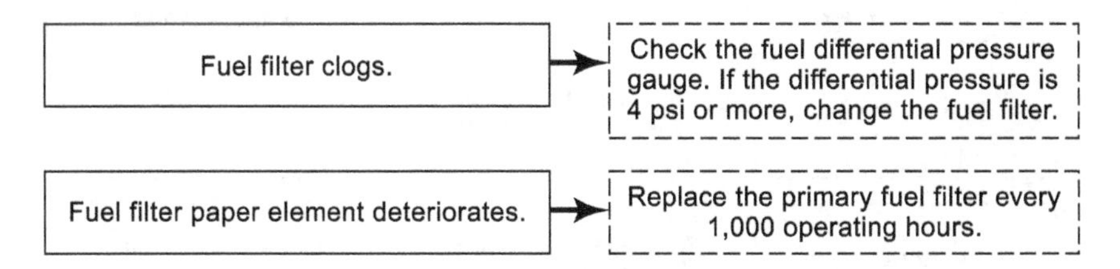

Figure 6.11 Two different failure management strategies for a fuel filter

As depicted in Figure 6.11, two different Failure Modes are required because there are two different failure management strategies required to manage the Failure Modes.

Ensure that the Failure Mode is a Cause of Failure and Not an Effect of a Failure Mode

When writing Failure Modes, care should be taken to ensure that the Failure Mode is actually a *cause* of failure and not a Failure Effect. This example shows once again why it is so important to have a trained facilitator lead the working group. Consider the following examples that are *not* Failure Modes.

- Bearing seizes.
- Generator coupling breaks.
- Brake drums score.
- Inadequate fuel is delivered to the engine.

For each of these examples, the facilitator would ask the question: *What specifically causes X to Y?* This process is specified for each example in Figure 6.12.

Ensure a Failure Mode is Reverse-Engineered for All Operating Context Concerns.

For example, assume an Operating Context concern documented by the working group is:

There is lack of oversight in Building 123 because the building is not managed full time. This leads to lack of ownership and communication which adversely affects the equipment.

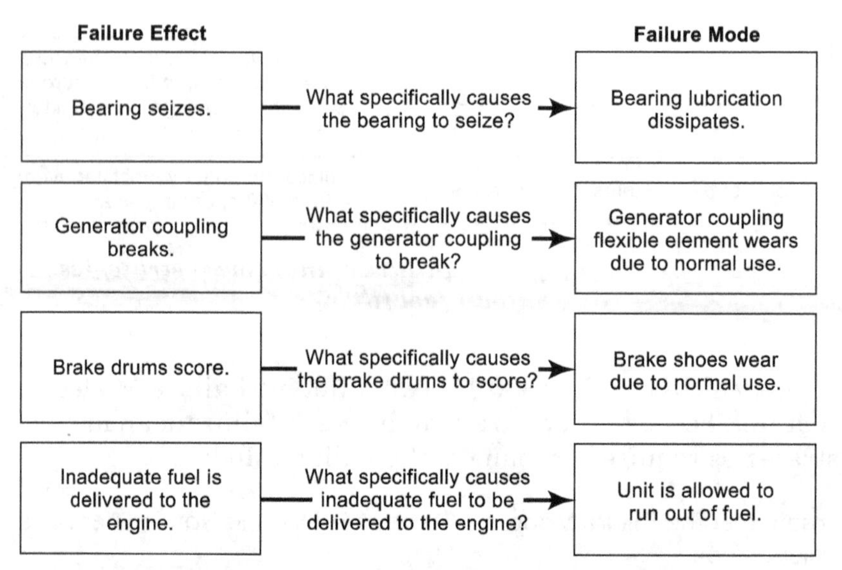

Figure 6.12 Failure Effects versus Failure Modes

This concern should be reversed engineered into the following Failure Mode and included in the RCM analysis.

Building 123 does not have a dedicated full-time building manager.

Writing the Failure Mode allows the issue to be analyzed using steps 4-7 of the RCM process.

Summary

A Failure Mode is what *specifically causes a Functional Failure*. Identifying Failure Modes is the third step in the RCM process. When identifying Failure Modes, the working group documents what could *cause* the Functional Failures listed in the Information Worksheet. Typical equipment Failure Modes such as wear and corrosion should be included in the Information Worksheet. However, because RCM isn't just about maintenance, Failure Modes regarding other issues such as human error and technical publications should also be included, allowing a host of other solutions to be derived.

Failure Effects

Writing Failure Effects is the fourth step in the RCM process. This chapter explores:

- What a Failure Effect is
- Recording Failure Effects on the Information Worksheet
- How to compose Failure Effects
- Writing Failure Effects for Protective Devices

Failure Effect
A story of what would happen if nothing were done to predict or prevent the Failure Mode.

7.1 What Is a Failure Effect?

A Failure Effect is a story of what would happen if nothing were done to predict or prevent the Failure Mode. Failure Effects should be written in enough detail so that the working group can assess the consequences of the Failure Mode.

7.2 Failure Effects and the Information Worksheet

Failure Effects are recorded on the Information Worksheet, as depicted in Figure 7.1. There is only one Failure Effect written for each Failure Mode. The Failure Effect does not get its own identifier. It shares the identifier with the Failure Mode (e.g., 1A1).

Information Worksheet

	Function		Functional Failure		Failure Mode	Failure Effect
1	Primary Function	A	Total Functional Failure	1	Failure Mode	Failure Effect
		B	Partial Functional Failure	1	Failure Mode	Failure Effect
2	Secondary Function	A	Total Functional Failure	1	Failure Mode	Failure Effect
		B	Partial Functional Failure	1	Failure Mode	Failure Effect
3	Secondary Function	A	Total Functional Failure	1	Failure Mode	Failure Effect

Figure 7.1 Failure Effects are documented on the Information Worksheet

7.3 Composing Failure Effects

Because RCM is a zero-based process, Failure Effects are written assuming nothing is being done to predict or prevent the Failure Mode.

Because RCM is a zero-based process, Failure Effects are written assuming nothing is being done to predict or prevent the Failure Mode. That is, Failure Effects are written assuming no proactive maintenance or other proactive strategy is being done. This is often a counter-intuitive process because equipment experts are diligent about taking care of their equipment. Therefore, they sometimes struggle having to write what would happen if they did nothing. This is another reason why it is so important to have a trained facilitator lead the working group. The facilitator coaches working group members so that Failure Effects are written properly.

Another important point about Failure Effects is that the worst case scenario is always documented. Varying degrees of Failure Effects often result from one Failure Mode. Documenting the worst case scenario ensures that the most conservative consequence is assessed. For example, if the braking system in a vehicle fails, the vehicle may be in an area where the car can be brought to a stop using the emergency brake without harm to the passengers or other people in the surrounding area. However, the braking system could fail in busy traffic such that the vehicle cannot be stopped before causing a fatal accident. The fatal accident is the worst case scenario.

Failure Effects

Failure Effects are written for each Failure Mode and should include the following.

- **A Description of the Failure Process from the Occurrence of the Failure Mode to the Functional Failure**

Include early evidence of the failure process such as phenomena that cannot be detected by the operating crew. This evidence can include vibration signatures that are only detectable using sophisticated monitoring devices or debris that accumulates in fluids such as hydraulic fluid, engine oil, and coolant.

Include evidence of the failure process that is detectable by the operating crew such as vibration, noise, heat, smoke, fluid leaks, gauge indications, warning lights, and various alarms. Include anything that the operating crew can see, feel, smell, or hear.

- **Physical Evidence that the Failure has Occurred**

Include evidence that the failure has occurred, such as bearing seizes, pump stops, engine stops, engine does not crank, hydraulic oil leaks onto the surrounding area, brakes do not engage, winch stops and cable does not pay in or out, drain hole clogs, and belt breaks.

- **How it Adversely Affects Safety and/or the Environment**

Specifically detail *how* someone could be injured or what environmental law or standard could be breached. Do not simply state: *operator could get hurt.* Instead be more specific, such as *operator could be electrocuted.*

- **How It Affects Operational Capability**

Explain how the operational capability is affected. For example: production stops for up to three days at a cost of $150,000 while an alternate means of producing instrument air is put in place—1,000 feet of scrap are produced at a cost of $20,000.

- **Specific Operating Restrictions as a Result of the Failure**

Detail any operating restrictions as a result of the failure. For example, if a minor steam leak is detected on a boiler system during operation, steam can still be produced. However, for safety reasons, there may be a restriction mandating that the boiler be shut down until the leak is repaired.

- Secondary Damage

Detail any secondary damage that may be caused by the Failure Mode. For example, the pump stops and the shaft shears; unfiltered hydraulic fluid is circulated through the system causing abnormal wear on hydraulic system components.

- What Must be Done and How Long It Takes to Repair the Failure (Downtime and Time to Repair)

At the end of the Failure Effect, detail what must be done to repair the failure and how long the repair will take. Do not write a detailed repair procedure. Simply state what must be done to repair the failure (e.g., the pump is replaced).

When documenting how long the repair will take, use the term *downtime to repair* when the equipment is *down* as a result of the Failure Mode. Use the term *time to repair* when the equipment can still operate and a repair is made at the next available opportunity.

What the repair time includes should be clearly explained. Typically the repair time starts with the time the failure occurs and goes until the equipment is back up and running. This period includes notifying maintenance of the failure, troubleshooting, supply lead time, time to make the repair, and a functional check. Some organizations choose not to include supply lead time in the interval because supply lead times can vary greatly and are sometimes quite long. The elements included in the interval should be clearly outlined in the Operating Context under *Operating Context Notes*.

The following two examples illustrate properly written Failure Effects.

> **Example 1: Failure Mode—Feedwater Pump Bearing Lubrication Dissipates.** Lack of lubrication causes the bearing to wear abnormally. Vibration levels increase. Eventually, noise develops and just before failure, friction increases such that heat and smoke are generated. The bearing seizes, the pump stops, and feedwater is no longer supplied to the boiler. The water level in the boiler drops and is indicated on the water level sight gauge. If this change goes unnoticed, eventually inadequate water is

available to continue producing steam. The output steam pressure decreases such that less than 10 psi is delivered to the paper drying process. The drop in output pressure is indicated on the steam pressure gauge. When the steam supplied to the paper drying process falls below 140 psi, the low steam pressure alarm sounds. Up to 20,000 feet of paper are not thoroughly dried before the steam to the paper drying process can be stopped. The paper is scrapped for recycle and the feedwater pump motor is replaced. Downtime to repair—8 hours.

Example 2: Failure Mode—Steam Drum Access Panel Bolts Overtorqued Due to Lack of Torque Sequence in Tech Manual SR-52118 Bolt heads are overtorqued and are rounded off, leaving the bolt head unable to be gripped for removal. The seal under the access panel is improperly seated and eventually steam and water leak. Water slowly seeps onto the boiler room floor. Steam also leaks, which causes an unmistakable noise, but the steam pressure to the paper drying process does not fall below 140 psi. It is unlikely that personnel will be cut or burned by the steam. The boiler is shut down as per safety bulletin BRR-728, which interrupts operations. The steam drum access panel seal and bolts are replaced. Downtime to repair—up to 6 hours.

7.4 Writing Failure Effects for Protective Devices

As discussed in Section 4.3 of Chapter 4, protective devices are devices and systems intended to protect people, the asset, and the organization *in the event* that another failure occurs. That is, failure of the protective device only matters in the event that there is another failure. For this reason, Failure Effects for protective device Failure Modes always start with the following:

This Failure Mode only matters in the event that…

This concept is illustrated in the following example.

Example: Failure Mode—Low Pressure Steam Alarm Switch System Fails Open This Failure Mode only matters in the event that boiler steam pressure supplied to the paper drying process falls below 140 psi. The low steam pressure alarm does not sound, the paper is not thoroughly dried, and up to 150,000 feet of paper are scrapped for recycle at a cost of $50,000.

Summary

Writing Failure Effects is the fourth step in the RCM process. A Failure Effect is a story of what would happen if nothing were done to predict or prevent the Failure Mode. Failure Effects should be written in enough detail so the working group can appropriately assess consequences. One Failure Effect is written for each Failure Mode and is recorded on the Information Worksheet.

Failure Consequences

Assessing Failure Consequences is the fifth step in the RCM process. This chapter discusses:

- What a Failure Consequence is
- An introduction to the Decision Diagram
- How to classify Failure Modes as Evident or Hidden
- How to record Evident and Hidden Failure Modes on the Decision Diagram
- Assessing the four Failure Consequences: Safety, Environmental, Operational, and Non-Operational

8.1 What Is a Failure Consequence?

The fifth Step in the RCM process is to assess consequences. A Failure Consequence identifies how each Failure Mode matters. In the context of RCM, four consequences are assessed: Safety, Environmental, Operational, and Non-Operational.

A Failure Consequence

Identifies how each Failure Mode matters.

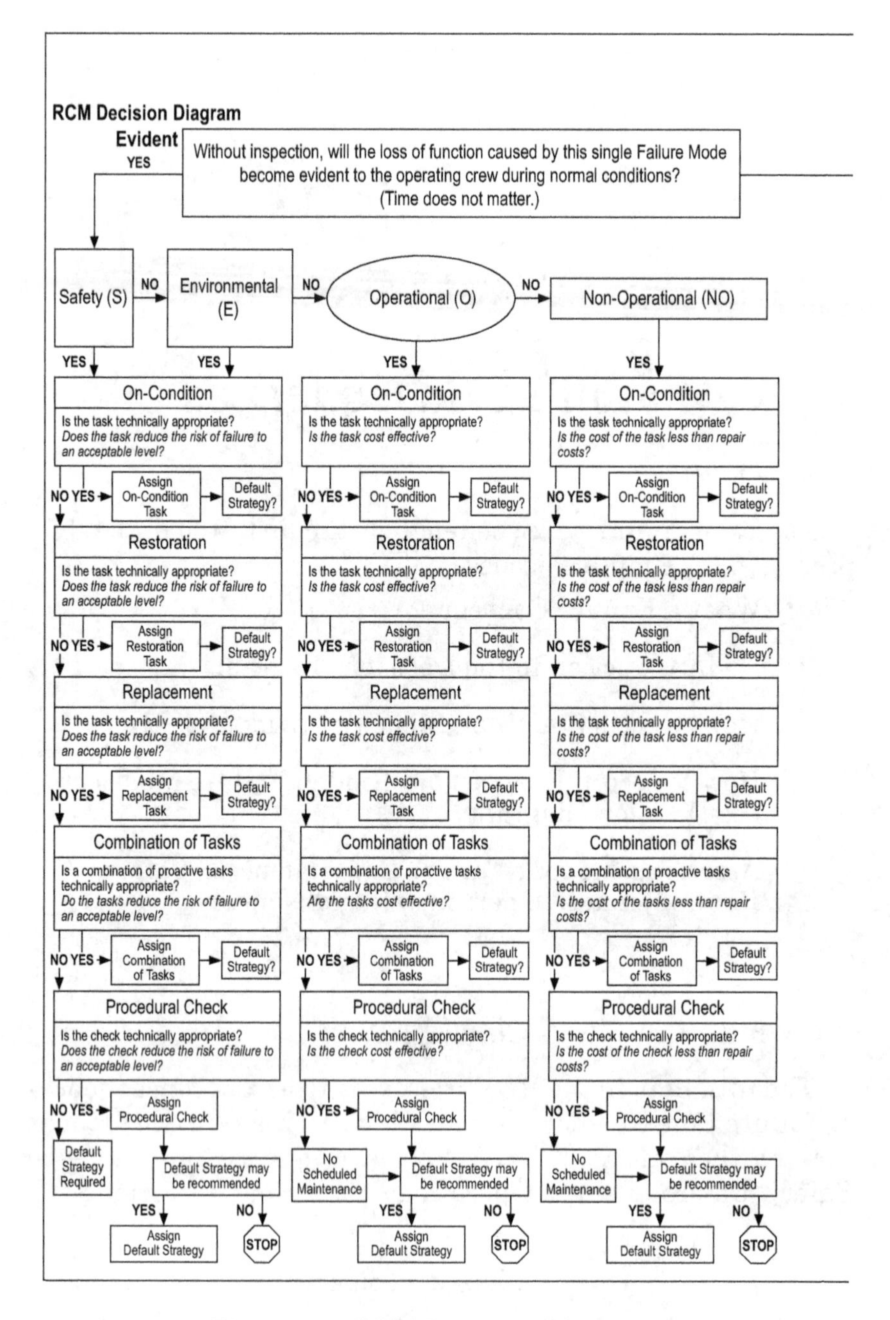

Figure 8.1 RCM Decision Diagram

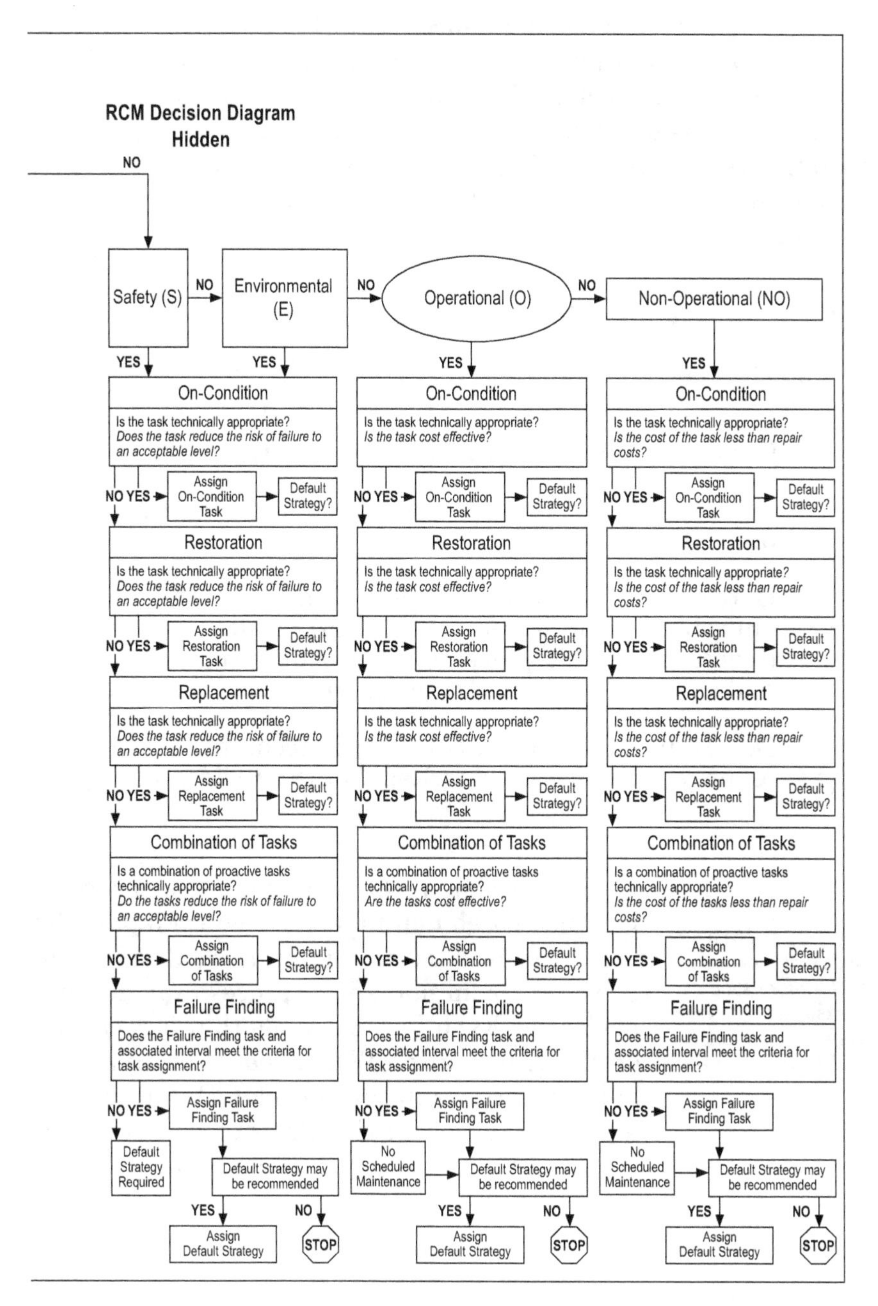

Figure 8.1 Continued

However, before consequences can be assessed, Failure Modes must be classified as Evident or Hidden. The Decision Diagram is used to classify Failure Modes as Evident or Hidden. The Decision Diagram is also used to complete steps five through seven of the RCM process: 5) assessing consequences, 6) identifying proactive maintenance, and 7) assigning default strategies.

8.2 Introduction to the Decision Diagram

The Decision Diagram is illustrated in Figure 8.1.
The Decision Diagram is an algorithm used to:

- Classify Failure Modes as Evident or Hidden
- Assess Failure Consequences (Step 5 of the RCM process)
- Determine if any proactive maintenance is technically appropriate and worth doing: On-Condition, Restoration, or Replacement tasks or a Combination of tasks (Step 6 of the RCM process)
- Determine if a Default Strategy is recommended to manage a Failure Mode (Step 7 of the RCM process)

Each Failure Mode that is recorded on the Information Worksheet is analyzed by being put through the Decision Diagram.

The Decision Diagram may appear complex but its elements, when reviewed piece by piece (as is done in Chapters 8, 9, and 10), present a common sense process for making decisions. Each Failure Mode that is recorded on the Information Worksheet is analyzed by being put through the Decision Diagram. The questions posed in the Decision Diagram allow logical solutions to be formulated.

8.3 Classifying Failure Modes as Evident or Hidden

The answer to the first question on the Decision Diagram determines whether a Failure Mode is considered *evident* or *hid-*

den. In order to determine whether a Failure Mode is evident or hidden, the first question on the Decision Diagram, as shown in Figure 8.2, is answered by the working group.

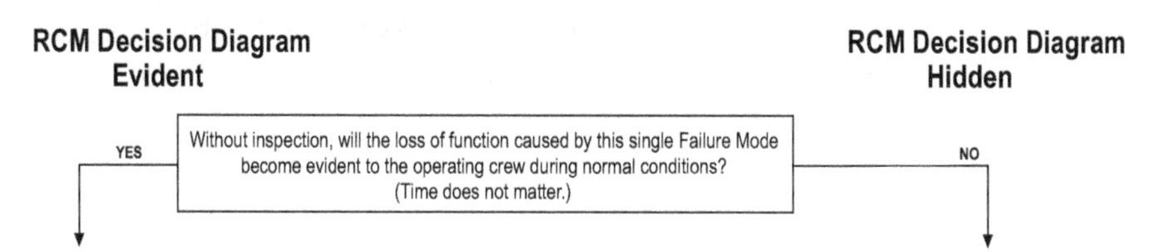

Figure 8.2 Determining if a Failure Mode is considered Evident or Hidden

Several phrases in Figure 8.2's question require discussion.

Failure Modes are analyzed assuming that nothing is being done to predict or prevent Failure Modes.

• Without inspection The phrase *without inspection* means that the question is answered assuming that no proactive maintenance is being performed. Proactive maintenance includes maintenance tasks as well as pre-operational checks, rounds, post-operational checks, walk-around checks, and other formal inspections. Assuming that no proactive maintenance is being performed is an element of RCM that allows a zero-based approach. That is, Failure Modes are analyzed assuming that nothing is being done to predict or prevent Failure Modes. In this way, unbiased solutions can be formulated.

Note: When operating equipment, there are certain actions performed that are not considered "proactive maintenance." In those cases, the working group can assume that the action is being accomplished when answering the first question on the Decision Diagram outlined in Figure 8.2. For example, assume a "start engine check" that indicates various parameters regarding the health of an engine is accomplished as a normal order of course each time an engine is started. It is considered "normal conditions." Therefore, when making RCM decisions, it is acceptable to assume that this check is being performed. Therefore, any indications or warnings that occur as a result of the check should be factored in when answering the first question on the Decision Diagram to determine if the Failure Mode is evident or hidden. An Operating Context note should always be included in the Operating Context when making any such assumptions.

For the "start engine check" example outlined here, the following Operating Context Note should be included in the Operating Context:

Failure Effects are written and Failure Consequences are assessed assuming that the operating crew performs a start engine check.

This note makes it clear that the start engine check is not considered a formal inspection, but rather a natural order of course when starting the engine. Therefore, if a failure occurred within the engine that, during startup, caused a warning light to illuminate, the light alerts the crew that a failure has occurred, which makes the loss of function *evident* to the operating crew.

Single Failure Mode The question in Figure 8.2 must be answered assuming that only the single Failure Mode being analyzed has occurred. All else within the system is considered to be working normally.

Normal conditions The phrase *normal conditions* implies that all associated equipment, systems, and operating context circumstances are normal. For example, when operating a personal vehicle, normal conditions include the oil pressure being within acceptable parameters, adequate fuel flow to the injectors, and an experienced driver at the wheel. These conditions are detailed in the Operating Context before an analysis begins.

Time does not matter When deciding if a Failure Mode is evident or hidden, the issue of time is not a factor. If the Failure Mode on its own becomes evident to the operating crew—even if it takes a year or longer to do so—the Failure Mode is still considered evident. For example, if the air conditioner in a vehicle used in Maine fails in October, it is likely that the vehicle owner would not become aware of it until the next summer when the driver switches on the air conditioner on the first warm day of the season. If this is the case, the Failure Mode is still evident because the driver (i.e., the operating crew) becomes aware of it without any other failures occurring.

• Final Note *Without inspection, would loss of function caused by this single Failure Mode become evident to the Operating Crew under normal conditions? (Time does not matter)*

The first question on the Decision Diagram does not ask if the operating crew knows exactly what *caused* the failure. The first question on the RCM Decision Diagram only asks if there is any phenomenon (e.g., excessive vibration, audible noise, engine quits, a warning light is illuminated) that is experienced by the operating crew to alert them that a failure is in the process of occurring or has already occurred. Suppose the boiler feedwater pump stops running. At the moment inadequate steam is delivered to the downstream process and the low steam pressure alarm sounds, the operating crew knows something is wrong but they don't know specifically what caused the pump to stop. The Failure Mode is designated as evident because it came to the operating crew's attention (by the low steam pressure alarm sounding) without any inspection.

Evident Failure Modes: Answering "Yes" to the First Question in the Decision Diagram

Without inspection, would loss of function caused by this single Failure Mode become evident to the Operating Crew under normal conditions? (Time does not matter.)

As depicted in Figure 8.3, answering *yes* to the first question on the Decision Diagram classifies the Failure Mode as Evident. In that case, the working group moves to the left on the Decision Diagram as shown in Figure 8.1 and the entire right side (the hidden side) of the Decision Diagram is immediately eliminated from analysis for that Failure Mode.

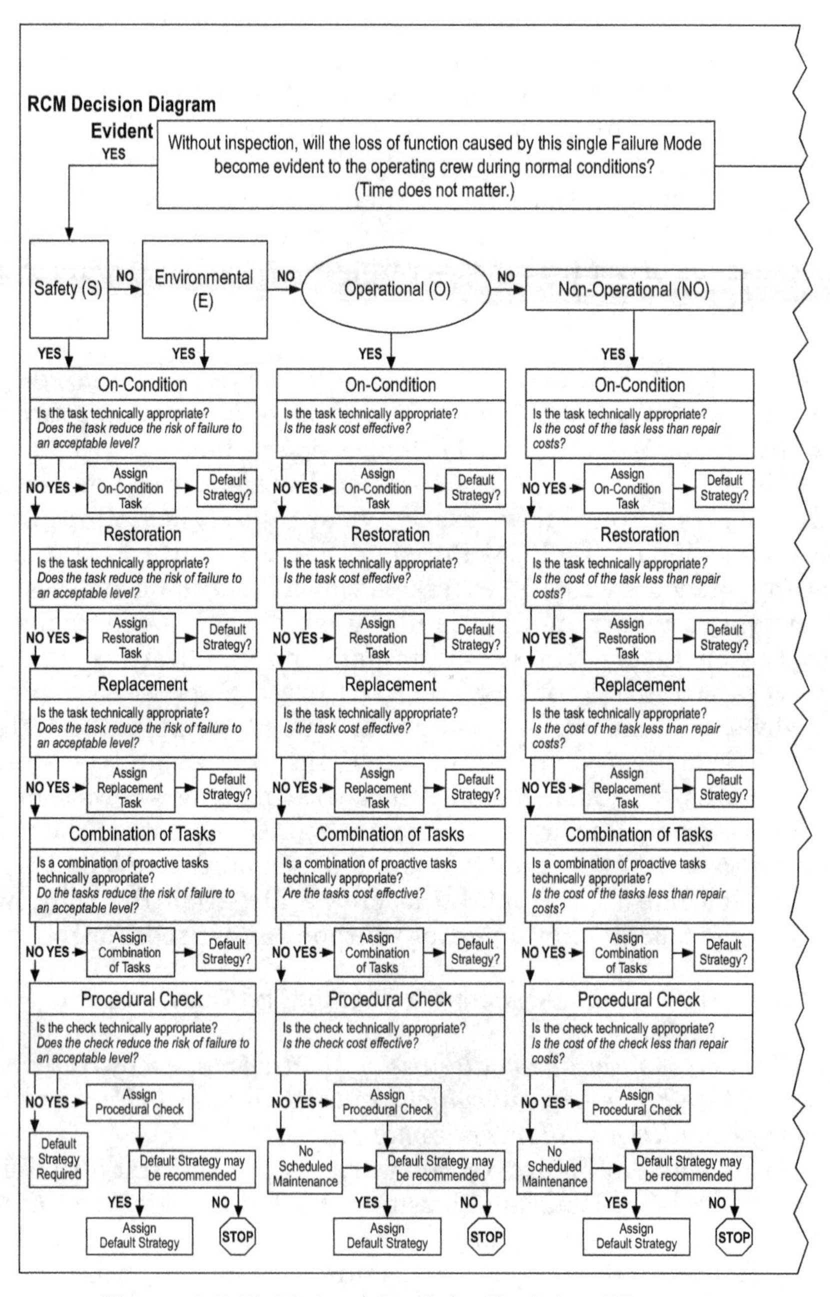

Figure 8.3 Evident side of the Decision Diagram

Consider the following examples.

Example 1

Failure Mode: Tow tractor primary fuel filter clogs.
Excerpt from Failure Effect: Eventually fuel flow to the engine is restricted. The engine runs rough and the engine shuts down due to lack of fuel.

The facilitator asks the working group the first question in the Decision Diagram: *Without inspection, would loss of function caused by this single Failure Mode become evident to the Operating Crew under normal conditions? (Time does not matter)*
Yes, the loss of function caused by the Failure Mode becomes evident to the operating crew when the engine shuts down. Therefore, the Failure Mode is classified as *evident.*

Example 2

Failure Mode: Any freon line fatigues due to normal use
Excerpt from Failure Effect: The Freon is depleted to the atmosphere. The temperature of the space being cooled increases and activity in the space is stopped.

The facilitator asks the working group the first question in the Decision Diagram: *Without inspection, would loss of function caused by this single Failure Mode become evident to the Operating Crew under normal conditions?*
Yes, the loss of function caused by the Failure Mode becomes evident to the operating crew when personnel inside the space being cooled report that it's getting hot and activity in the space is stopped. Therefore, the Failure Mode is classified as *evident.*

Hidden Failure Modes: Answering "No" to the First Question on the Decision Diagram

Without inspection, would loss of function caused by this single Failure Mode become evident to the Operating Crew under normal conditions? (Time does not matter.)

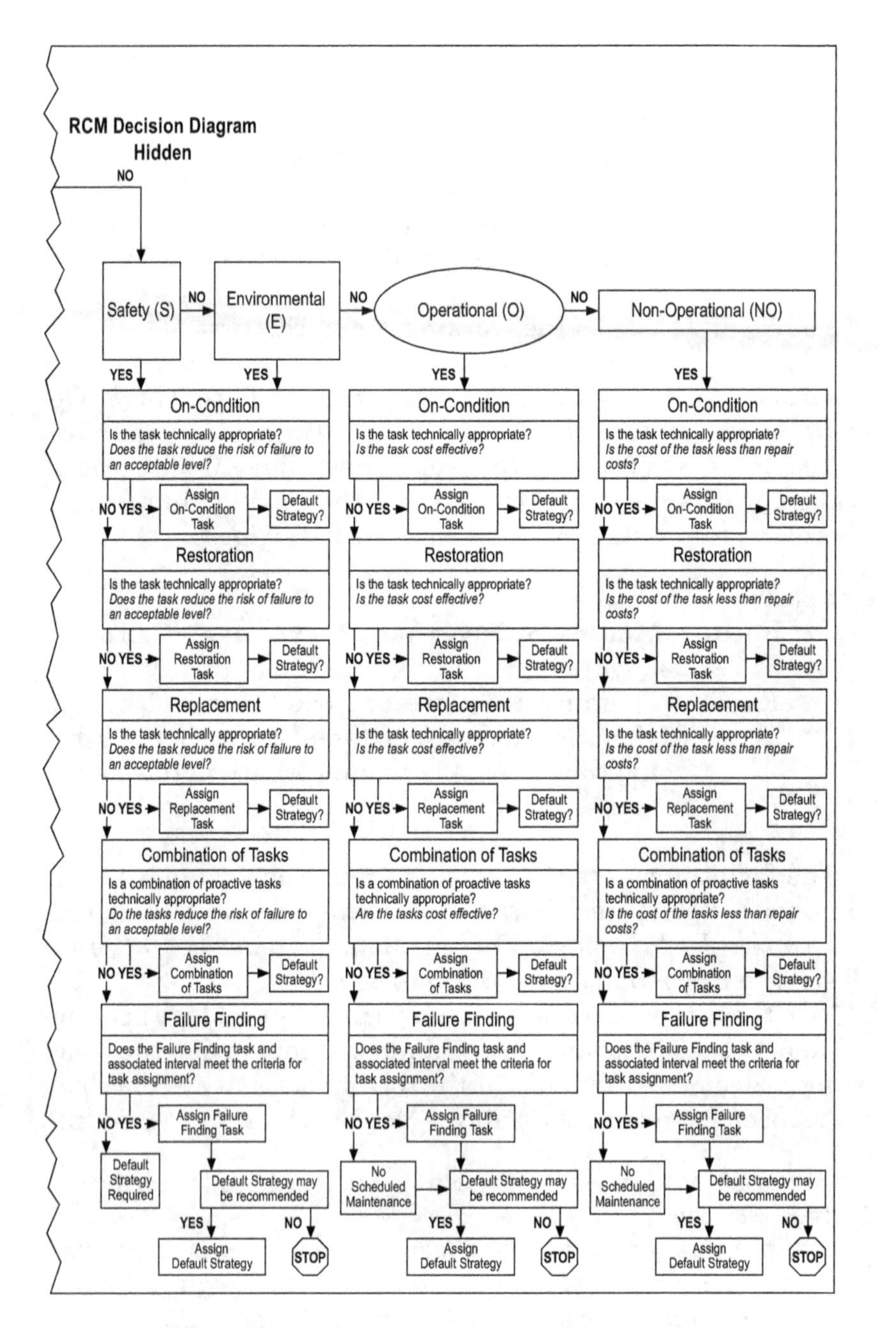

Figure 8.4 Hidden side of the Decision Diagram

As depicted in Figure 8.1, answering *no* to the first question on the Decision Diagram classifies the Failure Mode as Hidden. In that case, the working group moves to the right on the Decision Diagram and the entire left side (the evident side) of the Decision Diagram is immediately eliminated from analysis for that Failure Mode.

Consider the following examples.

Example 1

Failure Mode: Hydraulic system pressure safety valve fails closed.

Excerpt from Failure Effect: Valve is unable to relieve pressure in the event that the pressure exceeds 4,000 psi. In the event that the pressure exceeds 4,000 psi and the pressure safety valve is failed closed, hydraulic components rupture. Personnel in the surrounding area may be burned by the hot hydraulic fluid and the loose debris from the rupture.

The facilitator asks the working group the first question in the Decision Diagram: *Without inspection, would loss of function caused by this single Failure Mode become evident to the Operating Crew under normal conditions? (Time does not matter)*

No, the loss of function caused by the pressure safety valve that is failed closed does not become evident to the operating crew. The key in this example is that, under *normal conditions*, the operating pressure would never exceed 4,000 psi. Therefore, the pressure safety valve failing closed *on its own* would not become evident to the operating crew. It only matters *in the event* that the pressure exceeds 4,000 psi. Therefore, this Failure Mode is hidden. The pressure safety valve failing closed *and* the pressure exceeding 4,000 psi, in the context of RCM, is the *multiple failure*.

Example 2

Failure Mode: Shear pin is inadvertently replaced with a bolt.

Excerpt from Failure Effect: In the event that the recovery cable is overloaded, the cable snaps and personnel in the vicinity can be seriously injured by the whipping cable.

The facilitator asks the working group the first question on the Decision Diagram: *Without inspection, would loss of function caused by this single Failure Mode become evident to the Operating Crew under normal conditions? (Time does not matter.)*

In this case, no, the loss of function caused by the shear pin being replaced with a bolt does not become evident to the operating crew. Under *normal conditions*, the recovery cable would not be overloaded. Therefore, *on its own*, the shear pin being replaced with a bolt would not become evident to the operating crew. It only matters *in the event* that the recovery cable is overloaded. Therefore, the Failure Mode is hidden. The shear pin being replaced with a bolt *and* the recovery cable being overloaded is the *multiple failure.*

In summary, hidden Failure Modes have no direct adverse effects. The result is an increased exposure to the consequences of a *multiple failure.*

Recording Evident and Hidden Failure Modes

Once it is determined if a Failure Mode is evident or hidden, it is recorded on the RCM Decision Worksheet. The RCM Decision Worksheet is depicted in Figure 8.5. The first column header *Failure Mode* is where the Failure Mode designator is recorded (e.g., 1A1). The second column header is represented by an *E* which stands for *evident*. If the Failure Mode is evident, *Y* is recorded in the box. If the Failure Mode is hidden, *N* is recorded in the box. An example of Failure Mode 1A1, an evident Failure Mode, is recorded in Figure 8.5.

The following list of acronyms is used in the RCM Decision Worksheet. Each column is explained when the associated con-

RCM Decision Worksheet

Failure Mode			E	ES	EE	EO	ENO	HS	HE	HO	HNO	OC	Rst	Rpl	C	PC	FF	DS	Task	Initial Interval	Default Strategy
1	A	1	Y																		

Figure 8.5 RCM Decision Worksheet

cepts are discussed in this chapter and in Chapters 9 and 10.

ES: Evident Safety

EE: Evident Environmental

EO: Evident Operational

ENO: Evident Non-Operational

HS: Hidden Safety

HE: Hidden Environmental

HO: Hidden Operational

HNO: Hidden Non-Operational

OC: On-Condition task

RST: Restoration task

RPL: Replacement task

C: Combination of Tasks

PC: Procedural Check

FF: Failure Finding Task

DS: Default Strategy

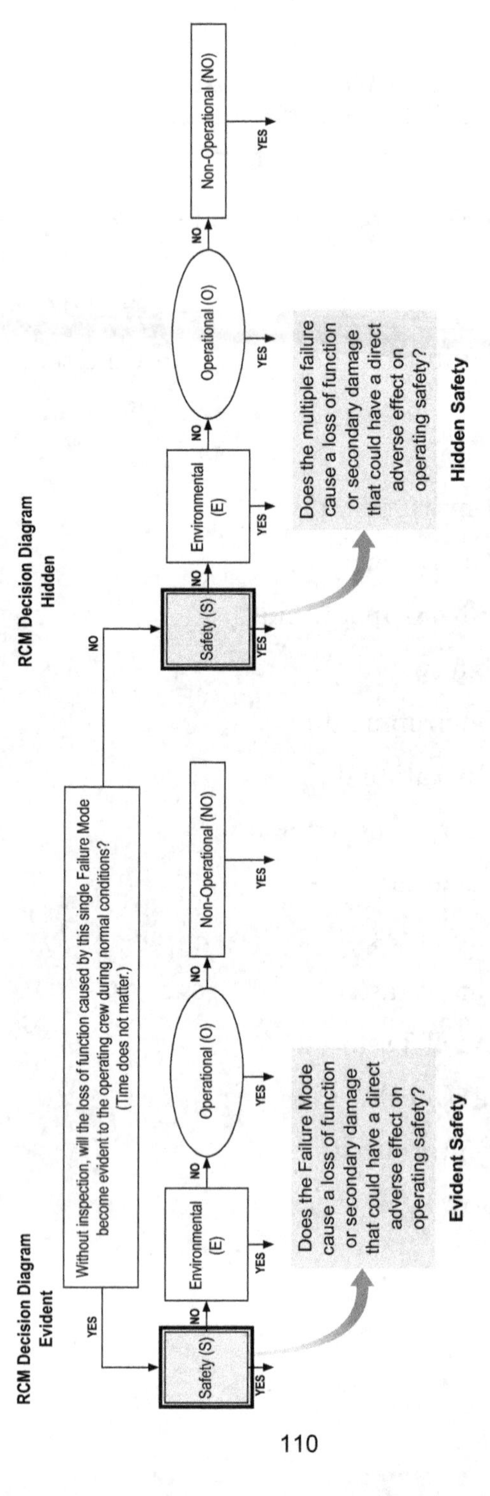

Figure 8.6 Assessing Safety Consequences using the RCM Decision Diagram

8.4 Identifying Failure Consequences

Once Failure Modes are classified as evident or hidden, the Decision Diagram is used to assess Failure Consequences.

Once Failure Modes are classified as evident or hidden, the Decision Diagram is used to assess Failure Consequences. Whether a Failure Mode has been classified as evident or hidden, one of four consequences must be assessed in the following order of priority: safety, environmental, operational, and non-operational.

Safety Consequences

Safety consequences are always assessed at the most conservative level. As depicted in Figure 8.6, the Decision Diagram poses questions for determining evident-safety and hidden-safety consequences.

Evident Safety Consequences

As depicted in Figure 8.6, when a Failure Mode is classified as evident, the working group moves to the left on the Decision Diagram and answers the question in the evident-safety consequence box. The answer to the question determines if the Failure Mode causes a loss of function or secondary damage that could have an adverse effect on operating safety. If anyone could be harmed or killed, obviously the Failure Mode has safety consequences. In this case, as depicted in Figure 8.7, *Y* (for yes) is placed in the Evident Safety (ES) column of the RCM Decision Worksheet. (Note that the columns HS, HE, HO, and HNO are inapplicable because the Failure Mode is evident.)

RCM Decision Worksheet

Failure Mode	E	ES	EE	EO	ENO	HS	HE	HO	HNO	OC	Rst	Rpl	C	PC	FF	DS	Task	Initial Interval	Default Strategy
1 A 1	Y	Y																	

Figure 8.7 Documenting an Evident-Safety Failure Mode on the RCM Decision Diagram

Hidden Safety Consequences

As depicted in Figure 8.6, when a Failure Mode is classified as hidden, the working group moves to the right on the Decision Diagram and answers the question in the hidden-safety consequence box. The answer to the question determines if the *multiple failure* causes a loss of function or secondary damage that could have an adverse effect on operating safety. This is a different question than what is posed for evident Failure Modes. For hidden Failure Modes, if anyone could be harmed or killed as a result of the *multiple failure*, then the multiple failure has safety consequences. Consider the following example regarding the hydraulic system pressure safety valve.

Example

Failure Mode identifier: 8A1

Failure Mode: Hydraulic system pressure safety valve fails closed.

Excerpt from Failure Effect: This Failure Mode only matters in the event that the hydraulic pressure exceeds 4,000 psi. In the event that the pressure exceeds 4,000 psi, hydraulic components rupture. Personnel in the surrounding area may be burned by the hot hydraulic fluid and the loose debris from the rupture.

In this example, the safety question on the Decision Diagram asks the following: *Does the multiple failure (the pressure safety valve fails closed* and *hydraulic system pressure exceeds 4,000 psi) cause a loss of function or secondary damage that could have a direct adverse effect on operating safety?* Of course, the answer is *Yes*.

In this case, as depicted in Figure 8.8, *N* (for no) is placed in the *E* column of the RCM Decision Worksheet signifying that the Failure Mode is hidden (not evident). Additionally, *Y* is placed in the Hidden Safety (HS) column of the RCM Decision Worksheet which indicates that the multiple failure has safety consequences. Note that the columns (ES, EE, EO, and ENO are inapplicable because the Failure Mode is hidden and thus they are skipped.)

If a Failure Mode or a multiple failure could have safety consequences, but the level of risk associated with it is acceptable, then a working group may choose to answer *No* to the question of safety consequences.

RCM Decision Worksheet

Failure Mode			E	ES	EE	EO	ENO	HS	HE	HO	HNO	OC	Rst	Rpl	C	PC	FF	DS	Task	Initial Interval	Default Strategy
8	A	1	N					Y													

Figure 8.8 Documenting Hidden Safety Consequences on the RCM Decision Diagram

Environmental Consequences

In order for a Failure Mode or a multiple failure to be assigned environmental consequences, its effect must breach an environmental law, standard, or regulation.

In order for a Failure Mode or a multiple failure to be assigned environmental consequences, its effect must breach an environmental law, standard, or regulation. If, for example, a gallon of oil is spilled as a result of a Failure Mode, but the oil is contained properly by hazardous material personnel, then the oil spill does *not* constitute environmental consequences. Just like safety consequences, environmental consequences are assessed at the most conservative level.

If a Failure Mode or a multiple failure does not have safety consequences, the working group moves to the right from the safety block on the Decision Diagram in order to consider environmental consequences. The questions posed for both evident and hidden environmental consequences are depicted in Figure 8.9.

Evident and Hidden Environmental Consequences

For evident Failure Modes, the consequences of the Failure Mode are assessed. For hidden Failure Modes, the consequence of the multiple failure is assessed.

Similar to safety consequences, there is a difference in the evident-environmental and hidden-environmental boxes on the RCM Decision Diagram. For evident Failure Modes, the consequences of the *Failure Mode* are assessed. For hidden Failure Modes, the consequence of the *multiple failure* is assessed. If

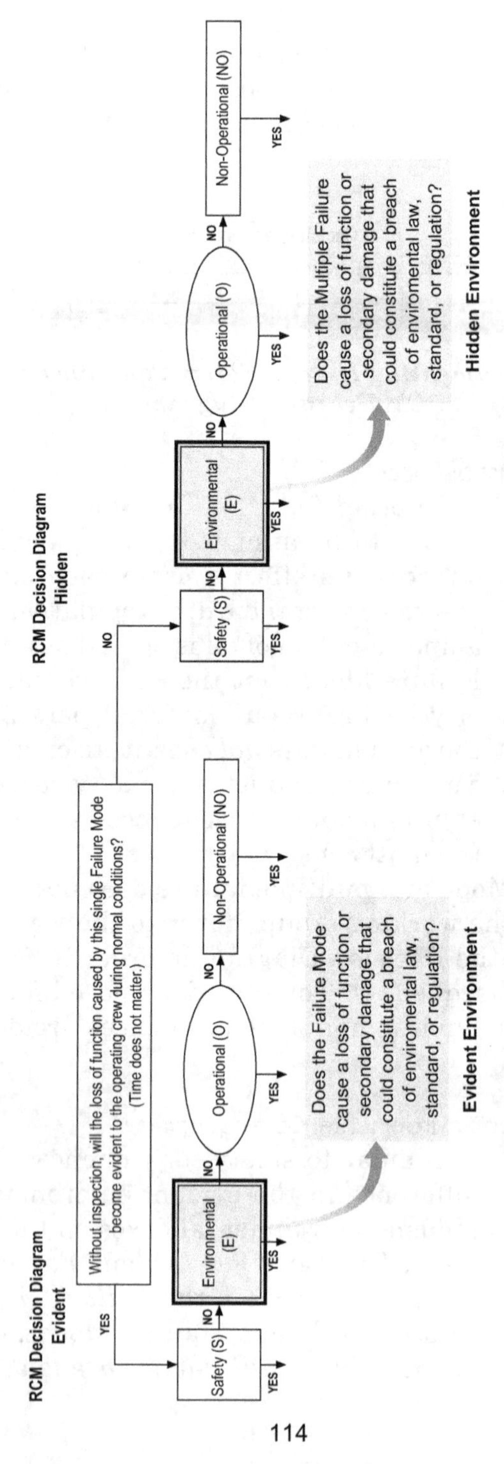

Figure 8.9 Assessing Environmental Consequences using the RCM Decision Diagram

an evident Failure Mode has environmental consequences, *Y* is placed in the Evident Environmental (EE) column of the RCM Decision Worksheet, as depicted in Figure 8.10 as Failure Mode 3A1. Conversely, for a hidden Failure Mode, if the multiple failure has environmental consequences, *Y* is placed in the Hidden Environmental (HE) column, as depicted in Figure 8.10 for Failure Mode 4A1.

RCM Decision Worksheet

Failure Mode	E	ES	EE	EO	ENO	HS	HE	HO	HNO	OC	Rst	Rpl	C	PC	FF	DS	Task	Initial Interval	Default Strategy
3 A 1	Y	N	Y																
4 A 1	N					N	Y												

Figure 8.10 Documenting Evident and Hidden Environmental Consequences on the RCM Decision Diagram

Operational Consequences

In order for a Failure Mode or a multiple failure to be assigned operational consequences, its effect must adversely affect operational capability. For example, operations may be disrupted, manufacturing may be delayed, or a mission may be aborted or canceled altogether. Additionally, the Failure Mode or multiple failure may not delay or stop operations, but it may increase operating cost. For example, if an oxygen sensor in a vehicle fails, fuel consumption is increased, but the vehicle can still fulfill its function. It just costs more to do so. In the context of RCM, a failed oxygen sensor has operational consequences because it increases operating cost. In the context of RCM, note that Failure Modes or multiple failures that may cause an inconvenience, but do not adversely affect operations, *do not* have operational consequences.

As depicted in Figure 8.11, if a Failure Mode or a multiple failure does not have safety or environmental consequences, the working group moves to the right from the Safety and Environmental blocks on the Decision Diagram in order to consider operational consequences. The questions posed for both

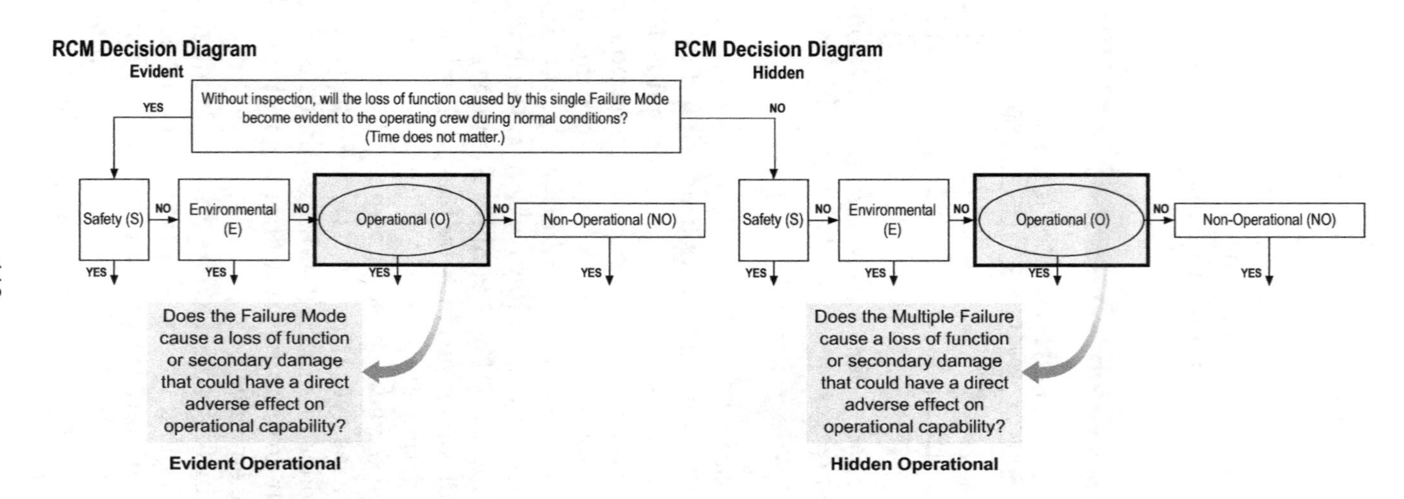

Figure 8.11 Assessing Operational Consequences using the RCM Decision Diagram

evident and hidden operational consequences are depicted in Figure 8.11.

Evident and Hidden Operational Consequences

Similar to safety and environmental consequences, there is a difference in the evident-operational and hidden-operational boxes on the RCM Decision Diagram. For evident Failure Modes, the consequences of the *Failure Mode* are assessed. For hidden Failure Modes, the consequences of the *multiple failures* are assessed. If an evident Failure Mode has operational consequences, a *Y* is placed in the Evident Operational (EO) column of the RCM Decision Worksheet, as depicted in Figure 8.12 for Failure Mode 5A1. Conversely, for a hidden Failure Mode, if the multiple failure has operational consequences, a *Y* is placed in the Hidden Operational (HO) column, as depicted in Figure 8.12 for Failure Mode 6A1.

RCM Decision Worksheet

Failure Mode			E	ES	EE	EO	ENO	HS	HE	HO	HNO	OC	Rst	Rpl	C	PC	FF	DS	Task	Initial Interval	Default Strategy
7	A	1	Y	N	N	N	Y														
8	A	1	N					N	N	N	Y										

Figure 8.12 Documenting Evident and Hidden Operational Consequences on the RCM Decision Diagram

The Epitome of Operational Consequences

The epitome of assessing operational consequences occurred on February 20, 2005. British Airways Flight 268, a Boeing 747, had just taken off from Los Angeles International Airport and was bound for London's Heathrow Airport. Moments after takeoff, flames were noticed coming from one of the engines. The pilots successfully shut down the affected engine and notified British Airways headquarters of the incident. British Airways management directed Flight 268 to carry on with the flight to London. Doug Brown, senior manager of British Airway's 747 fleet said: "The decision to continue flying was a customer service issue. The plane is as safe on three engines as it is on four and it can fly on two." British Airways quickly assessed the consequences of the failure. If they delayed or cancelled the flight, it could have cost British Airways up to several hundred thousand dollars in passenger compensation because of a recently passed European regulation regarding long flight delays or cancellations. The choice was made to avoid the operational consequences (cost of the delayed or cancelled flight) and continue on to London because it was assessed that the aircraft could safely do so.

Non-Operational Consequences

In order for a Failure Mode or a multiple failure to be assigned non-operational consequences, its effect must only involve the cost of repair. That is, there is no adverse effect on safety, the environment, or operations.

Consider the following example.

Example

Failure Mode: Pintle light circuit fails open.

Excerpt from Failure Effect: The operator uses an alternate means to illuminate the pintle hook and towing operations continue without delay.

The Failure Mode *Pintle light circuit fails open* does not adversely affect safety, the environment, or operations. Therefore, its consequences are non-operational. As depicted in Figure 8.13, if a Failure Mode or a multiple failure does not have safety, environmental, or operational consequences, the working group moves to the right from the Safety, Environmental, and Operational blocks on the Decision Diagram to the Non-Operational block.

If an evident Failure Mode has non-operational consequences, a *Y* is placed in the Evident Non-Operational (ENO) column of the RCM Decision Worksheet, as depicted in Figure 8.14 for Failure Mode 7A1. Conversely, for a hidden Failure Mode, if the multiple failure has non-operational consequences, a *Y* is placed in the Hidden Non-Operational (HNO) column, as depicted in Figure 8.14 for Failure Mode 8A1. However, it is extremely rare that a multiple failure has non-operational consequences. It begs the question: If the consequences are non-op-

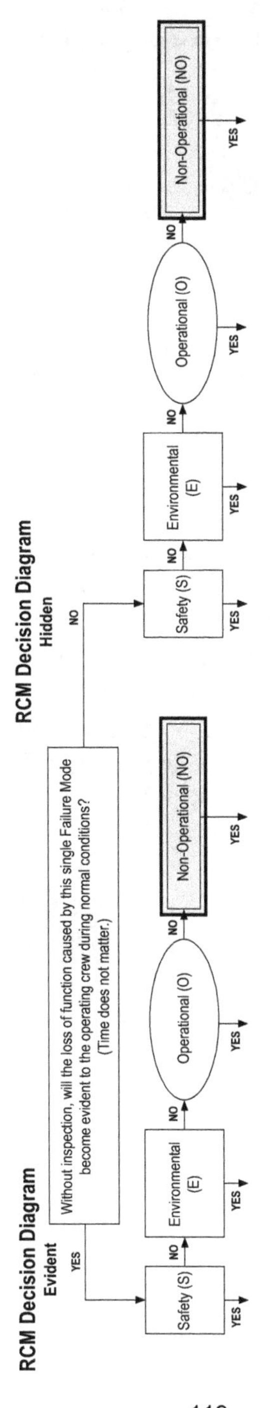

Figure 8.13 The Decision Diagram and Non-Operational Consequences

erational, why would a protective device be built into the system in the first place?

RCM Decision Worksheet

Failure Mode			E	ES	EE	EO	ENO	HS	HE	HO	HNO	OC	Rst	Rpl	C	PC	FF	DS	Task	Initial Interval	Default Strategy
7	A	1	Y	N	N	N	Y														
8	A	1	N					N	N	N	Y										

Figure 8.14 Documenting Evident and Hidden Non-Operational Consequences on the RCM Decision Diagram

Summary

Assessing Failure Consequences is the fifth step in the RCM process. A Failure Consequence identifies how each failure matters. However, before consequences can be assessed, Failure Modes must be classified as Hidden or Evident. The answer to the first question in the RCM Decision Diagram determines whether a Failure Mode is considered *evident* or *hidden*. Then the RCM Decision Diagram is used to assess one of four consequences: Safety, Environmental, Operational, and Non-Operational.

Nine

Proactive Maintenance and Intervals

Assigning proactive maintenance tasks and associated intervals is the sixth step in the RCM process. This chapter discusses:

- What constitutes proactive maintenance in the context of RCM
- Two criteria required for assigning a scheduled maintenance task
- Scheduled Restoration, Replacement, On-Condition tasks, Combination of Tasks and associated intervals
- How to document proactive tasks on the RCM Decision Worksheet
- Synchronizing initial task intervals

9.1 Proactive Maintenance in the Context of RCM

As depicted in Figure 9.1, there are two types of proactive maintenance: preventive maintenance and On-Condition maintenance.

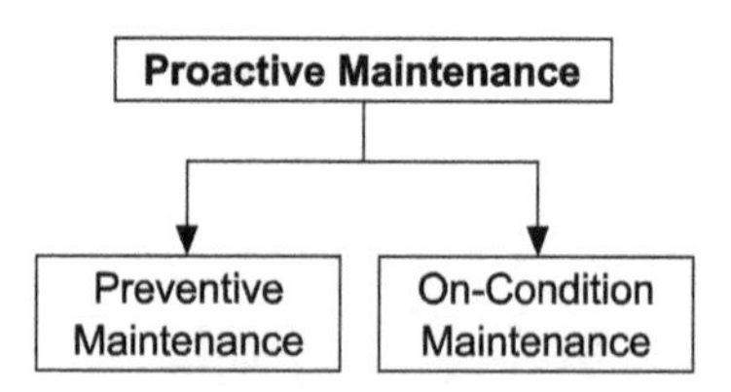

Figure 9.1 Two Types of Proactive Maintenance

In the context of RCM, preventive maintenance includes only scheduled Restoration and scheduled Replacement tasks.

In the asset management community, the term *preventive maintenance* is widely used to mean any kind of scheduled maintenance that is performed. However, in the context of RCM, preventive maintenance includes only scheduled Restoration and scheduled Replacement tasks, as depicted in Figure 9.2.

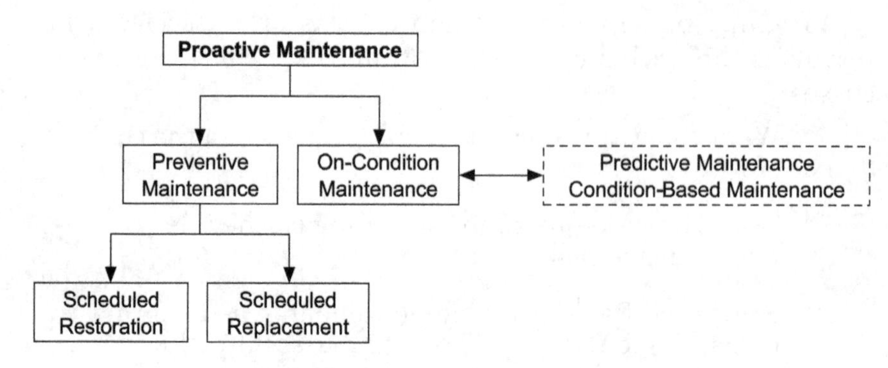

Figure 9.2 Preventive and On-Condition Maintenance

The other type of proactive maintenance is On-Condition maintenance. As depicted in Figure 9.2, other terms used for On-Condition maintenance are predictive maintenance and Condition-Based maintenance (CBM) and can be used interchangeably. For clarity, the term *On-Condition maintenance* is used throughout this book.

9.2 Criteria for Assigning a Proactive Maintenance Task

In order to schedule a proactive maintenance task in the context of RCM, two criteria must be satisfied. The proactive task must be:

1. Technically appropriate
2. Worth doing

The RCM Decision Diagram is used to determine if a proactive maintenance task meets two criteria: technically appropriate and worth doing.

The RCM Decision Diagram is used to determine if a proactive maintenance task meets these criteria. The following discussion details Restoration, Replacement, and On-Condition tasks in addition to what *technically appropriate* and *worth doing* means for each type of task.

9.3 Scheduled Restoration and Scheduled Replacement Tasks

The following discussion details scheduled Restoration and scheduled Replacement tasks.

Scheduled Restoration Task

A scheduled restoration task reworks or restores an item's failure resistance to an acceptable level without considering the item's condition at the time of the task.

An example of a scheduled Restoration task is removing a vehicle's coolant, recycling it, and replacing it in the vehicle every 100,000 miles.

Scheduled Replacement Task

A scheduled replacement task replaces an item without considering the item's condition at the time of the task.

An example of a scheduled Replacement task is replacing an automobile's timing belt at 105,000 miles.

In the context of RCM, scheduled Restoration and Replacement tasks are considered preventive maintenance because action is taken *before* failure is allowed to occur, thereby *preventing* failure. Scheduled Restoration and Replacement tasks are performed *without considering the condition of the item* at the time of the task. In other words, the item isn't inspected first to determine if preventive maintenance is necessary. Scheduled Rest-oration and Replacement tasks are performed at fixed intervals.

Scheduled Restoration or Replacement Tasks and the Decision Diagram

Scheduled Restoration and scheduled Replacement tasks are represented on the RCM Decision Diagram in the second and third rows, as outlined in Figure 9.3.

The questions embodied in the Decision Diagram within the Restoration and Replacement blocks are answered in order to determine if a scheduled Restoration or Replacement task is technically appropriate and worth doing.

Determining if a Scheduled Restoration or Replacement Task is Technically Appropriate

One of the most important elements that makes a scheduled Restoration or Replacement task technically appropriate is if there is a useful life. The useful life is the age at which a significant increase in the conditional probability of failure occurs, as depicted in Figure 9.4.

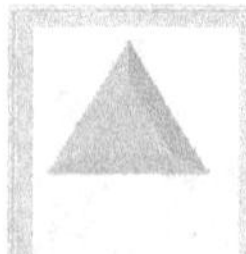

Useful Life

The age at which a significant increase in the conditional probability of failure occurs

That is, the likelihood of the Failure Mode occurring that the task is intended to prevent increases with operating age, as depicted in Figure 9.4.

Therefore, if a Failure Mode behaves according to this failure pattern, it makes sense that just before the end of the useful life is reached, a scheduled Restoration or Replacement task is performed in order to prevent failure from occurring. It is technically appropriate to do so if the conditional probability of failure increases at a certain age.

Note on conditional probability of failure: The *conditional probability* of failure is different than the *probability* of failure. Conditional probability of failure is the probability that a failure will occur at a specific "age" once the item has survived to that age. For example, a woman may have a 1 in 28 lifetime probability of dying of cancer. However, this is an average lifetime probability. Her chances of dying of cancer increase as she ages, so her risk at 40 is very different than her risk at 80. Just as a human's probability of dying from various illnesses increases *assuming a human being reaches a certain age*, so too do some Failure Modes. For example, the longer engine oil remains in service, the more likely that water intrudes and additives are depleted. Therefore, the probability of failure increases the longer the oil is in service.

Useful life is not the same as mean time between failures (MTBF).

Note on MTBF: The useful life is not the same as mean time between failures (MTBF). MTBF by definition is an average. That is, some failures will occur before the established MTBF and some will occur after. If MTBF is used to determine intervals for Restoration and Replacement tasks, some failures will be prevented. However, some failures will occur before the task is performed. Furthermore, many items will be replaced earlier than necessary while the items still have useful life remaining. It is essential to understand the difference between useful life and average life in order to ensure that correct information is being used when determining Restoration and Replacement intervals. This is another reason why it is so important that a trained facilitator leads the working group.

Using the Decision Diagram to Assign Scheduled Replacement and Scheduled Restoration Tasks

Figure 9.5 illustrates the evident-operational scheduled Restoration and Replacement blocks from the RCM Decision Diagram.

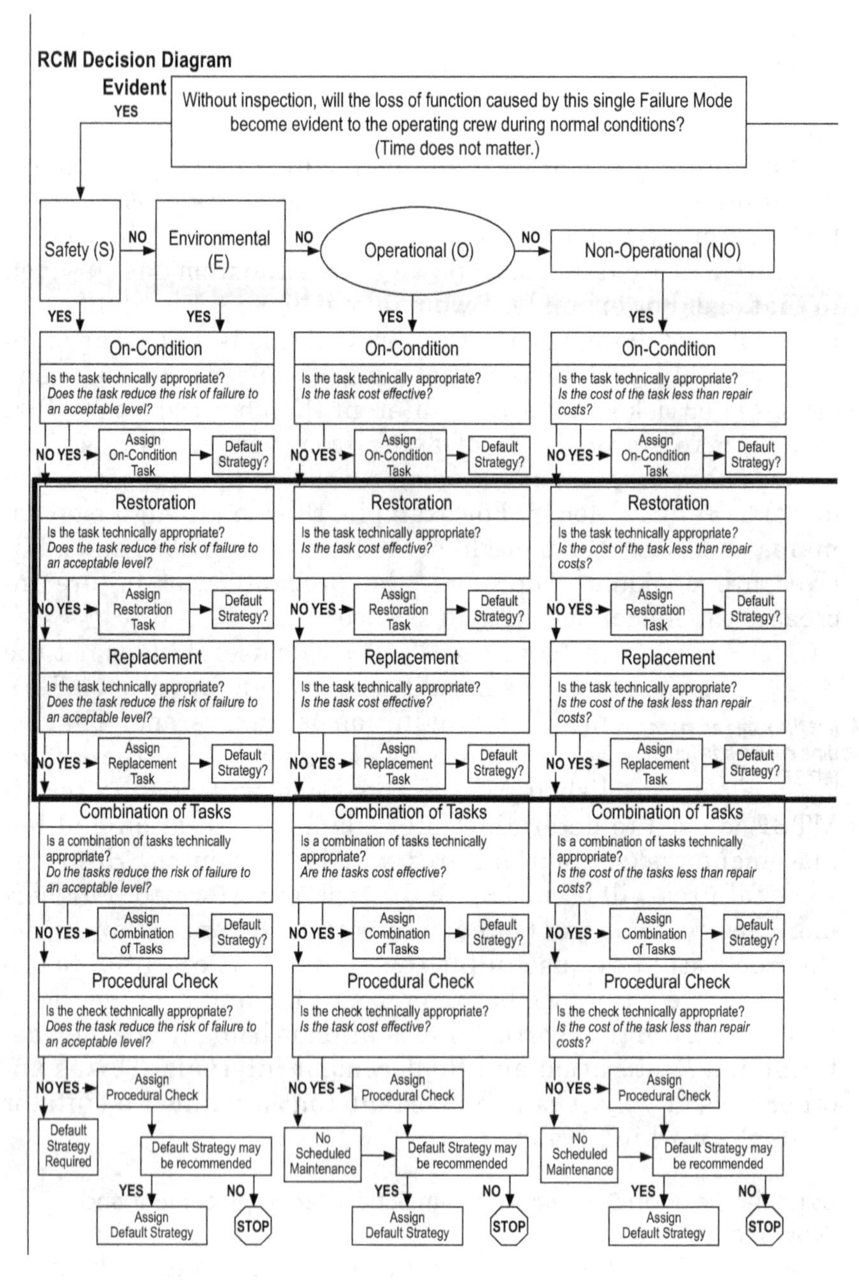

Figure 9.3 Restoration and Replacement Tasks represented on the Decision Diagram

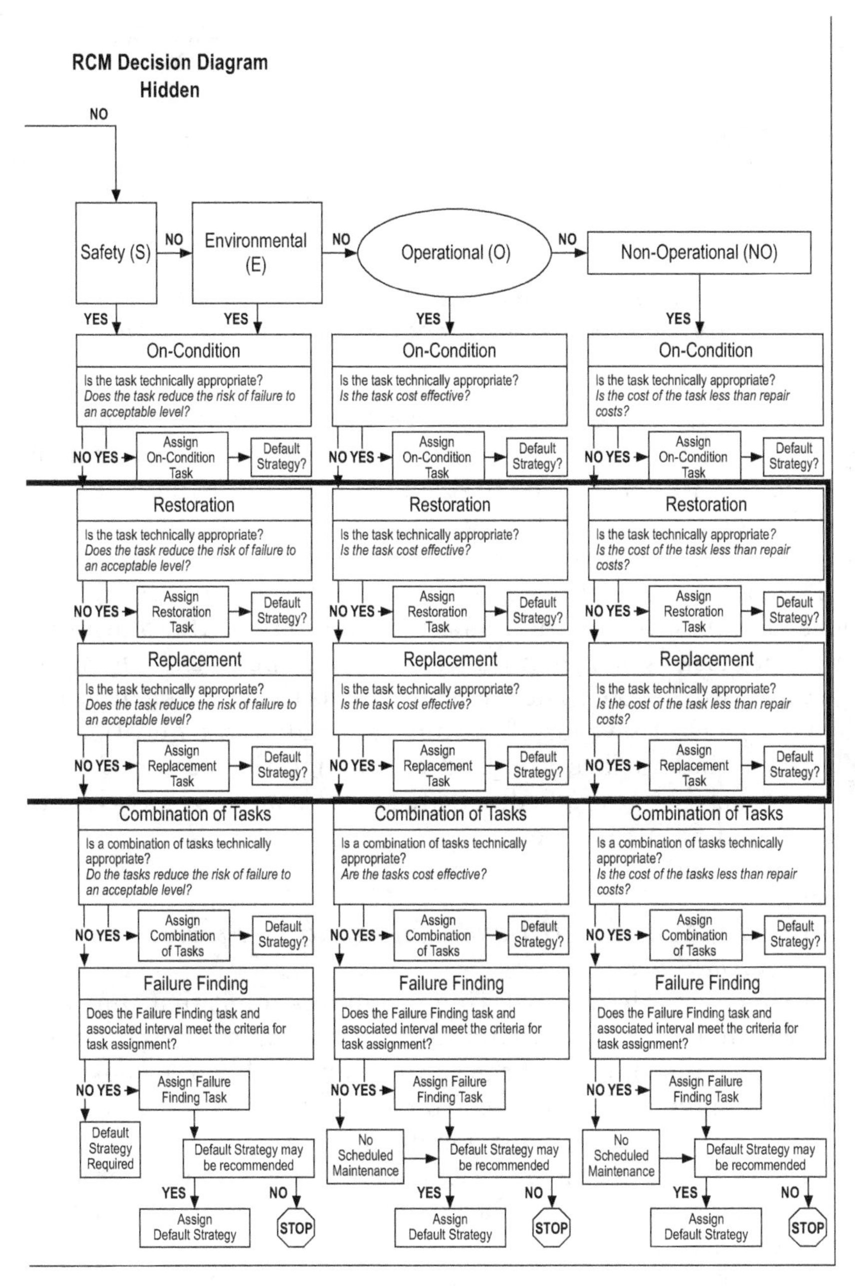

Figure 9.3 Continued

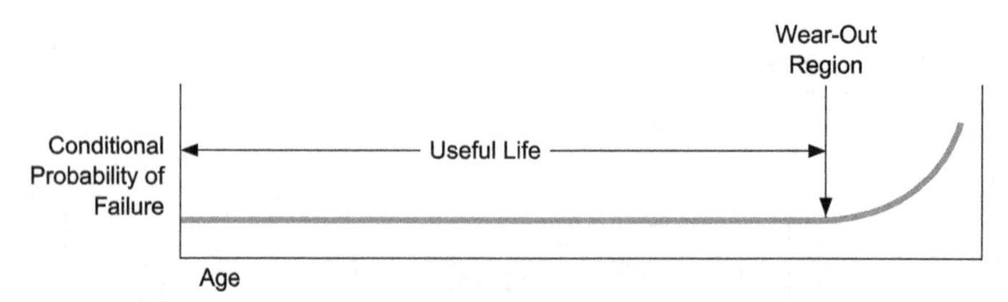

Figure 9.4 Failure Pattern depicting that the likelihood of the Failure Mode occurring increases with operating age

Notice that the text in the first set of questions within each block is normal font. However, the text in the second set of questions is italicized. The first set of questions establishes if a task is *technically appropriate*. The second set of questions establishes if the task is *worth doing*.

The facilitator asks the questions and the working group provides the answers. If the answer to any one of the questions within a block is *No*, then that task is not appropriate to manage the Failure Mode. The facilitator leads the working group to the next block. However, if the answer to all of the questions is *Yes*, then that task is assigned to manage the Failure Mode.

The following example illustrates these points for an evident-operational Failure Mode.

Example

Failure Mode: *2001 Honda CRV timing belt wears due to normal use.*

In this example, the working group has identified that the Failure Mode is evident and that it has operational consequences because failure of the timing belt would do serious secondary damage to the engine. From there, the facilitator led the working group to the Replacement block as shown in Figure 9.6, where the following questions were asked and answered in order to determine if a Replacement task is *technically appropriate and worth doing.*

Is there an identifiable wear-out age? (Does the probability of failure become significantly greater after a certain operating age?) *Yes*

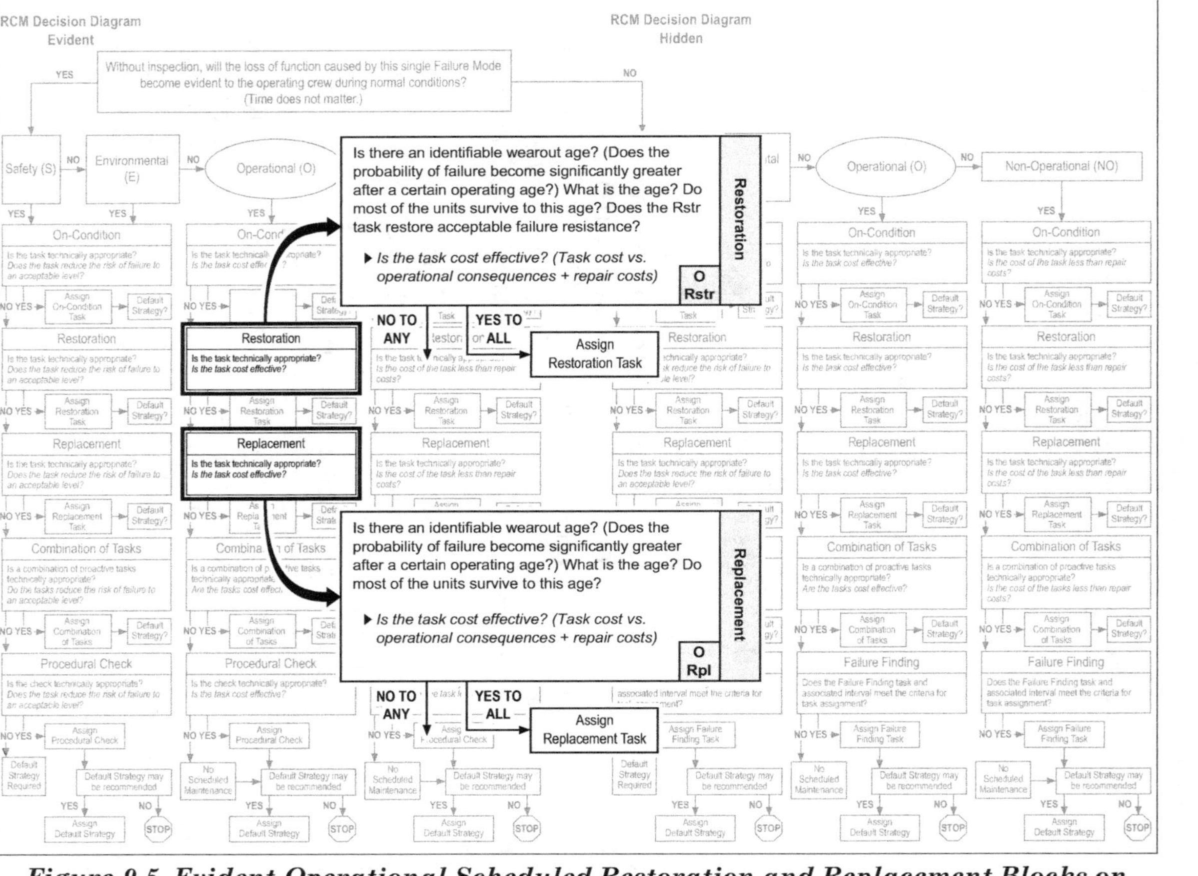

Figure 9.5 Evident-Operational Scheduled Restoration and Replacement Blocks on the RCM Decision Diagram

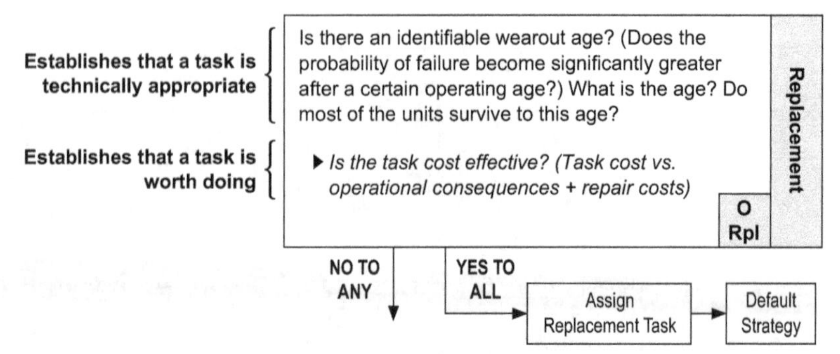

Figure 9.6 Establishing if a Replacement task is technically appropriate and worth doing

What is the age? *105,000 miles*
Do most of the units survive to this age? *Yes*

At this point, the working group has answered affirmatively to the first set of questions, thereby determining that it is technically appropriate to replace the timing belt at 105,000 miles.

Now the working group must determine if the task is cost effective by answering the last question. *Is the task cost effective? (Task cost versus operational consequences plus repair costs) Yes*

The working group has determined that it is cost effective to replace the timing belt versus letting it fail because the cost of the secondary damage far exceeds the cost of proactively replacing the timing belt. Because all of the questions have been answered affirmatively, the Replacement task is assigned.

Figure 9.7 illustrates how to document the above example on the RCM Decision Worksheet. The Failure Mode is evident so *Y* is placed in the E column. The working group determined that failure of the timing belt would have operational consequences so *N* is placed in the ES and EE blocks and *Y* is placed in the EO column. The working group answered *yes* to all the questions in the evident-operational block depicted in Figure 9.6, thereby establishing that the Replacement task is technically appropriate and worth doing. Therefore, *N* is placed in the On-Condition (OC) and Restoration (Rst) columns and *Y* is placed in the Replacement (Rpl) column. The task is written in the

RCM Decision Worksheet

Failure Mode			E	ES	EE	EO	ENO	HS	HE	HO	HNO	OC	Rst	Rpl	C	PC	FF	DS	Task	Initial Interval	Default Strategy
1	A	7	Y	N	N	Y						N	N	Y					Replace the timing belt.	105,000 miles	

Figure 9.7 Documenting a scheduled Replacement task on the RCM Decision Worksheet.

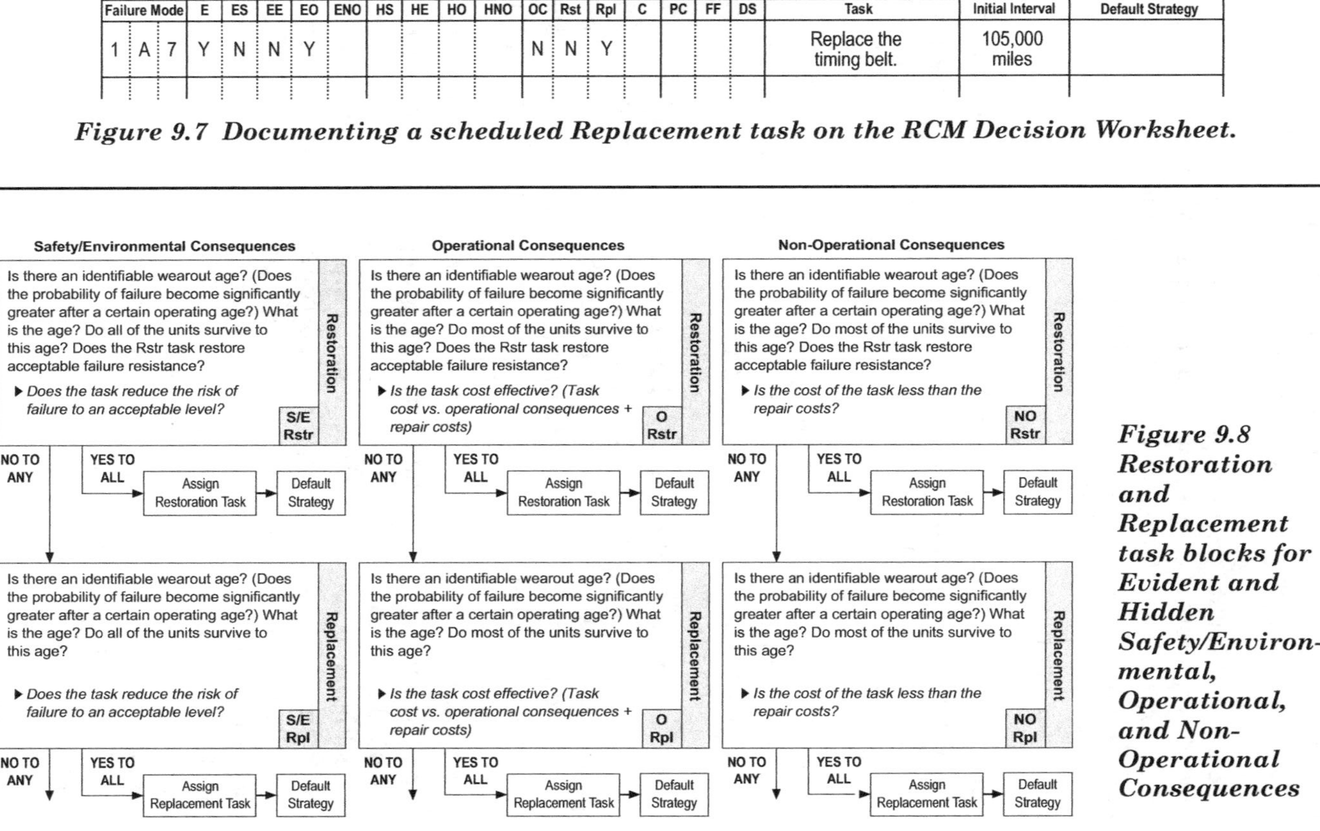

Figure 9.8 Restoration and Replacement task blocks for Evident and Hidden Safety/Environmental, Operational, and Non-Operational Consequences

task column and the initial interval of 105,000 miles is documented in the initial interval column.

Differences amongst Restoration and Replacement Task Blocks for Safety/Environmental, Operational, and Non-Operational Consequences

Figure 9.8 illustrates the hidden and evident Restoration and Replacement blocks from the RCM Decision Diagram for all consequences. The following discussion outlines the few differences amongst the blocks.

• **Determining if a Restoration Task is Technically Appropriate** Determining if a Restoration task is technically appropriate is almost identical to that of a Replacement task. The only difference is the addition of one question: *Does the Restoration task restore acceptable failure resistance?* So, for example, if automobile coolant is recycled, the working group must ensure that the recycled coolant is restored to a point that the coolant can protect engine components and prevent the engine from overheating.

• **Most versus All** As depicted in Figure 9.9, when dealing with safety or environmental consequences, the question on the Decision Diagram asks if *all* the units survive to this age. However, when dealing with operational or non-operational consequences, the question on the Decision Diagram asks if *most* of the units survive to this age.

Figure 9.10a shows most of the units surviving to the determined age versus *all* of the units surviving to that age, as represented in Figure 9.10b.

In Figure 9.10a, the random portion of the graph does not touch the x-axis illustrating an opportunity during the useful life for random failure to occur before reaching the end of the useful life. In Figure 9.10b, the random portion of the graph touches the x-axis illustrating that no failures will occur before reaching the end of the useful life. Of course, zero failures cannot be guaranteed. What is implied by "all," however, is that a safe-life limit has been established to show that no failures are expected to occur below the specified life limit.

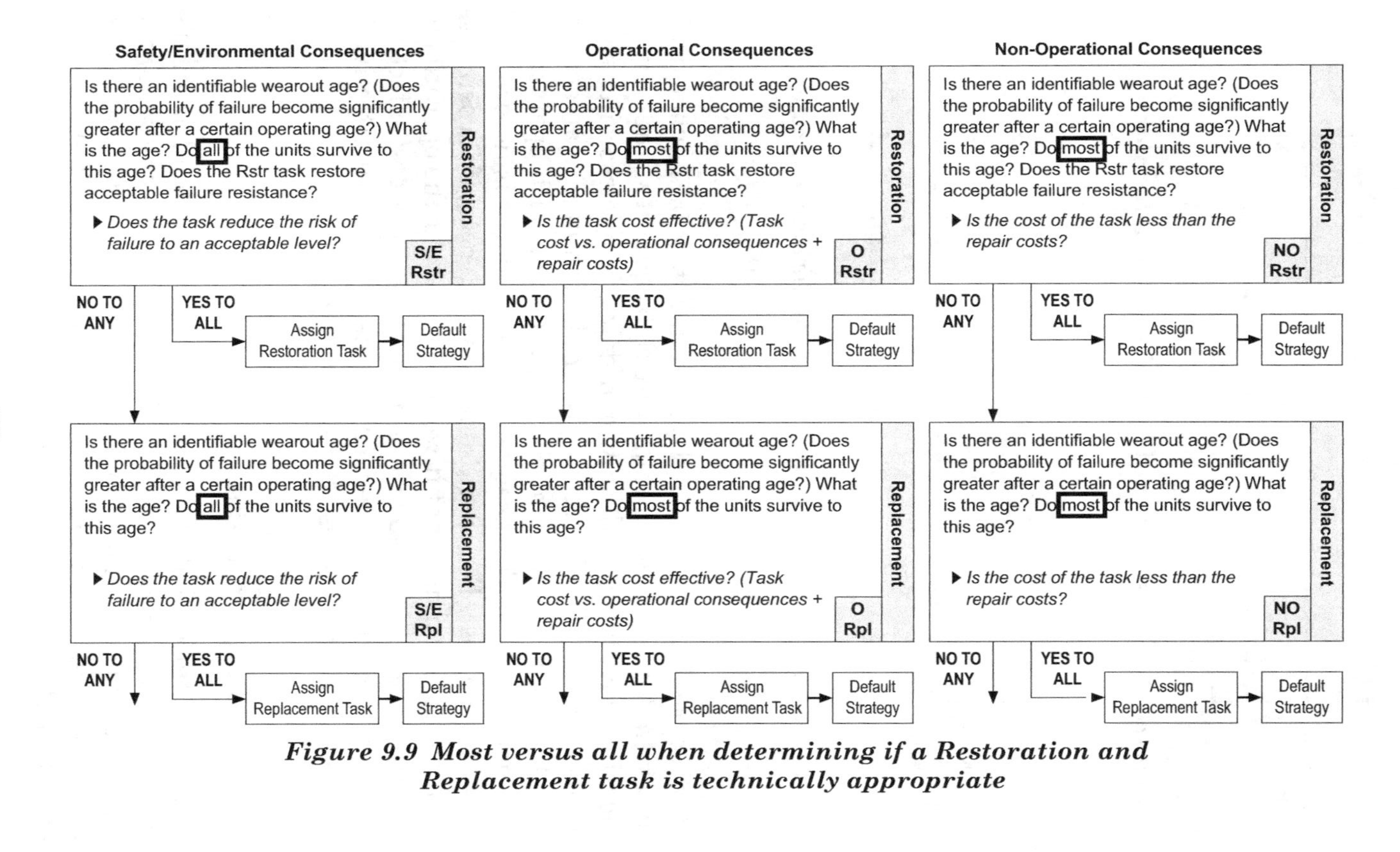

Figure 9.9 Most versus all when determining if a Restoration and Replacement task is technically appropriate

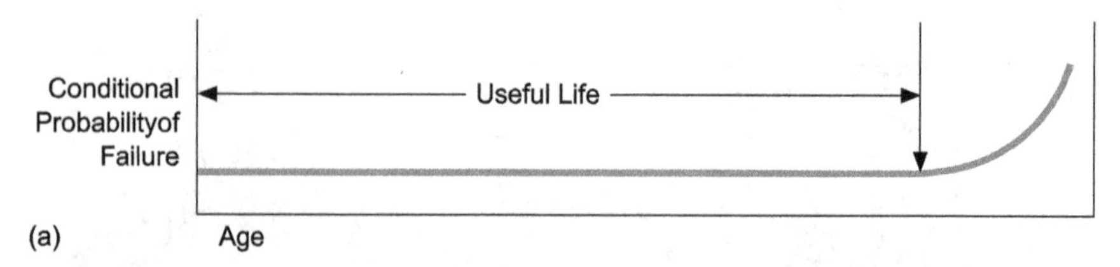

Figure 9.10a Most of the units survive to the determined age.

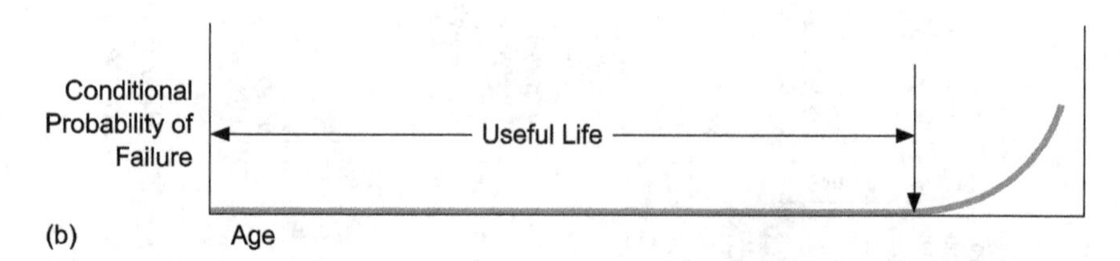

Figure 9.10b All of the units survive to the determined age.

Worth Doing: Risk versus Cost As depicted in Figure 9.11, when dealing with safety or environmental consequences, the way to measure whether a task is worth doing is by assessing risk. That is, the task has to reduce the risk of failure to an acceptable level. On the other hand, when dealing with operational and non-operational consequences, worth doing is determined by assessing cost. For operational consequences, if it costs less to do the task versus what it would cost to bear the operational consequences plus repair costs, then the task is cost effective. For non-operational consequences, the task is cost effective if the cost of the task is less than the repair costs.

9.4 On-Condition Tasks

Most Failure Modes give a warning that failure is in the process of occurring. That is, evidence of *impending failure* is generated such as vibration, heat, noise, wear, warning lights, and gauge indications. In the context of RCM, such evidence is called a *potential failure condition*.

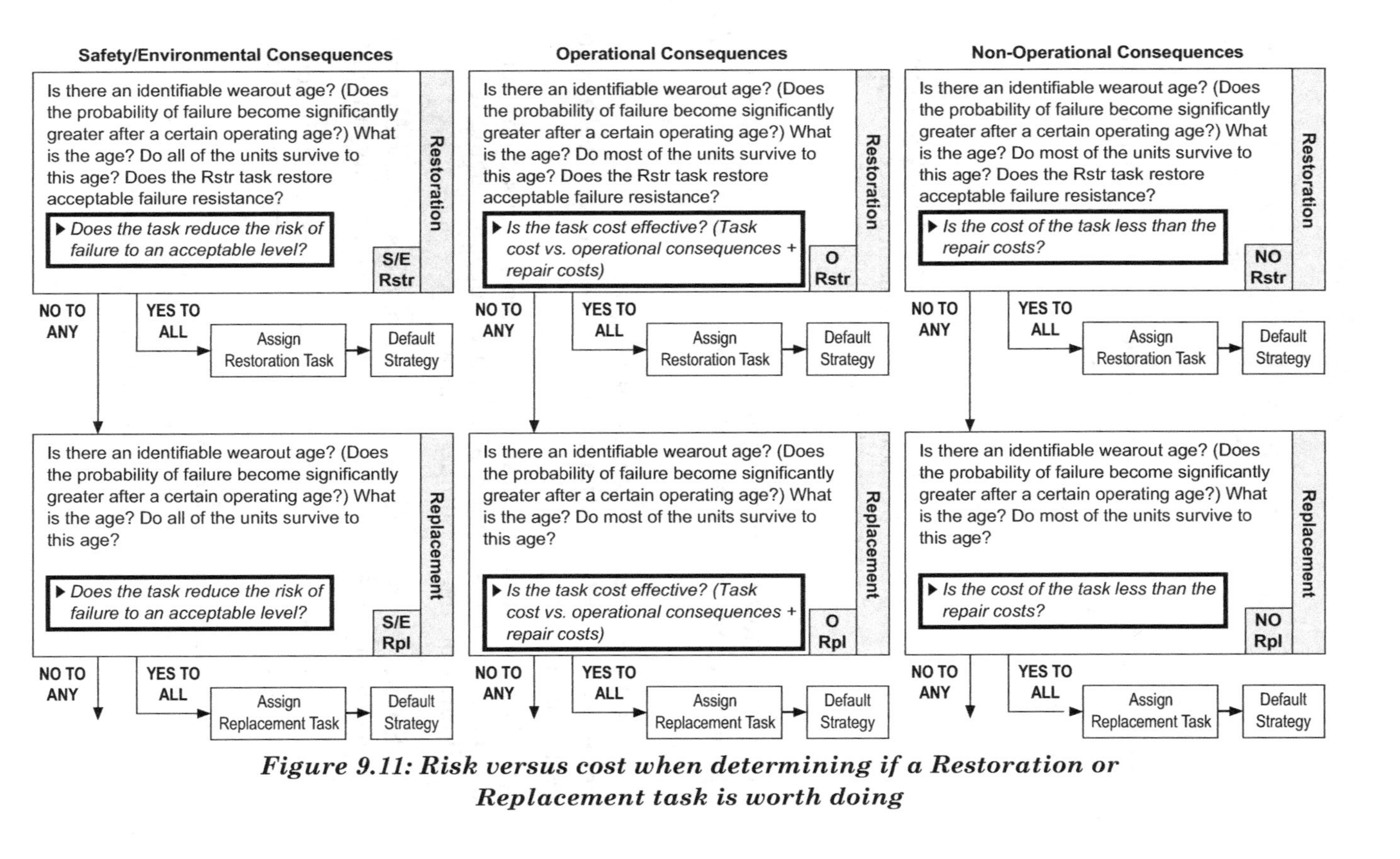

Figure 9.11: Risk versus cost when determining if a Restoration or Replacement task is worth doing

Potential Failure Condition

Evidence that a Failure Mode is in the process of occurring

An On-Condition task is performed at a defined interval to detect a potential failure condition so that maintenance can be performed *before* the failure occurs.

On-Condition task

A task performed at a defined interval to detect a potential failure condition so that maintenance can be performed *before* the failure occurs.

Therefore, maintenance is performed only on the evidence of need and an organization gets the maximum useful life out of a component.

How a Potential Failure Condition Can Be Detected

Potential failure conditions can be detected in several ways as detailed below.

- Using relatively simple techniques such as monitoring gauges, measuring brake linings, or monitoring for vibration using human senses
- Employing more technically-involved applications such as thermography, frequency analysis, or eddy current
- Continuous monitoring with devices installed directly on machinery (e.g., strain gauges and accelerometers)

On-Condition Tasks and the Decision Diagram

On-Condition tasks are represented on the RCM Decision Diagram in the first row as outlined in Figure 9.12.

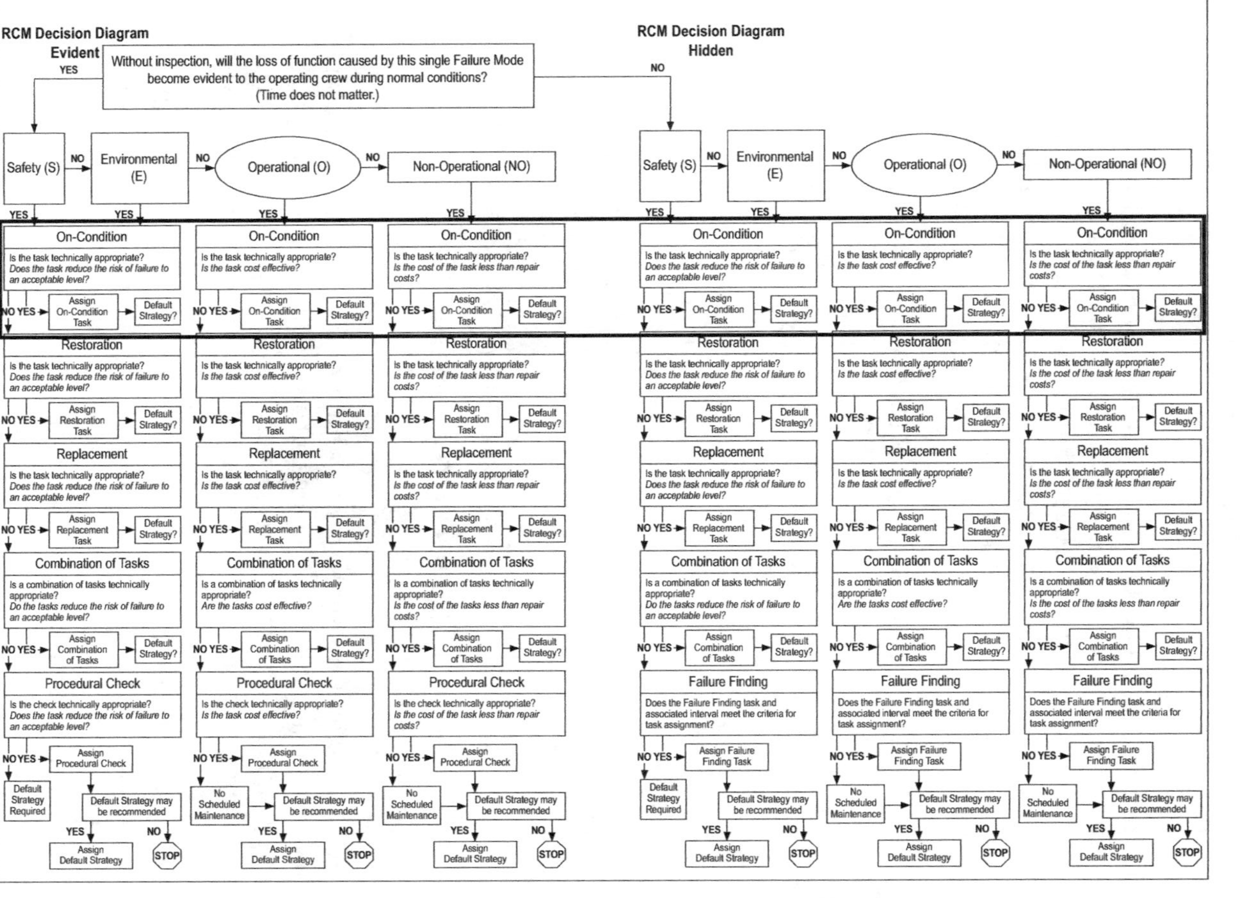

Figure 9.12: On-Condition Tasks represented on the Decision Diagram

The questions embodied in the Decision Diagram within the On-Condition blocks are answered in order to determine if an On-Condition task is technically appropriate and worth doing.

Determining If an On-Condition Task is Technically Appropriate

In order to determine if an On-Condition task is technically appropriate, potential failure conditions must be evaluated. The following discussion examines how On-Condition tasks are established to be technically appropriate.

P-F Curve

Figure 9.13 introduces the P-F Curve.

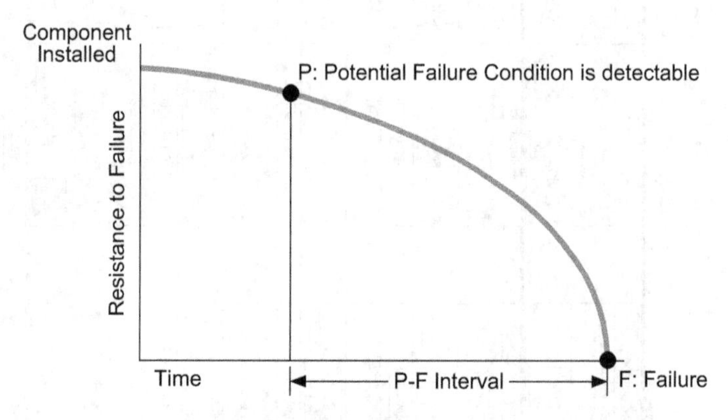

Figure 9.13 The P-F Curve

The x-axis represents time, which can be measured in any units such as calendar time, operating hours, miles, and cycles. The y-axis represents the resistance to failure. The time that a new component is installed is depicted at the top of the y-axis; it indicates the maximum amount of resistance to failure. As most components remain in service, the resistance to failure declines; eventually, signs of impending failure are exhibited. In other words, potential failure conditions develop. A potential failure condition is represented by *P* on the P-F curve. If the potential failure condition goes unheeded, eventually failure occurs. Failure is represented on the P-F curve as *F*. The P-F interval is the time from when the potential failure condition is *detectable* to the point that failure occurs.

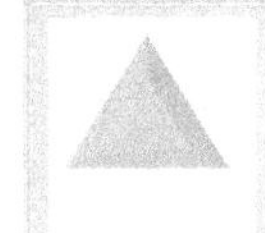

P-F Interval

The time from when the potential failure condition is *detectable* to the point that failure occurs.

On-Condition Task Intervals are Determined Based upon the P-F Interval

The manner in which On-Condition task intervals are set is a widely misunderstood concept. On-Condition task intervals are *not* based on MTBF—or, on average, how often the failure occurs. On-Condition task intervals are *not* based on the criticality of the failure. Instead, On-Condition task intervals are based on the P-F interval. A general rule of thumb is to perform the On-Condition task at half the P-F interval. However, this is merely a guide. As long as On-Condition task intervals are performed at intervals less than the P-F interval—and the net P-F interval, the minimum time remaining before failure occurs, leaves enough time to *manage the consequences of failure*—then the On-Condition task interval is acceptable.

On-Condition task intervals are based on the P-F interval.

Net P-F Interval

The minimum time remaining to take action in order to manage the consequences of failure.

The following example, depicted in Figure 9.14 for the Failure Mode *V-belt wears due to normal use*, demonstrates the P-F curve, the P-F interval, and the net P-F interval. It also indicates whether the net P-F interval is long enough to manage the consequences of failure.

Example Failure Mode: *V-belt wears due to normal use*

Before a V-belt breaks, it develops visual evidence of impending failure. The potential failure condition for the V-belt is evidence of wear (e.g., cracks and frays).

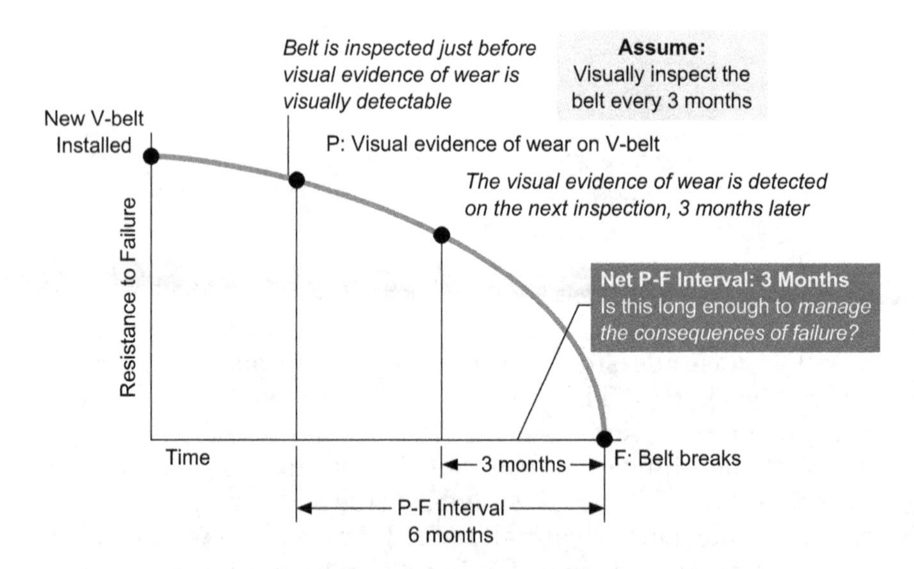

Figure 9.14: The P-F Curve example of a V-belt

P: *Visual evidence of wear on V-belt*

If the cracks and frays are ignored, the belt will eventually break.

F: *Belt breaks*

The working group reports that the P-F interval is 6 months. That is, the period from the time that visual evidence of wear on the belt is detectable to the point that the belt breaks is 6 months.

P-F interval: *6 months*

Assume that the P-F interval is halved and visual inspections of the belt are performed every 3 months.

On-Condition task interval *(how often the belt is in spected): 3 months*

Assume that the belt is inspected *just before* visual evidence of wear on the V-belt is detectable, as depicted by the arrow just in front of "P" on Figure 9.14. During the time between the next inspection, visual evidence of wear develops. However, the wear isn't discovered until the next inspection, which is three months later. That means that there is a minimum of three months remaining before the belt breaks. That is, the net P-F interval is 3 months.

Net P-F interval: *3 months*

Now the working group must determine if the P-F interval is long enough to accommodate an inspection every three months and if the net P-F interval is long enough to *manage the consequences of failure.*

Scenario 1: Assume that the asset with the V-belt is shut down only every 6 months. In this scenario, the item can't be monitored at intervals less than the P-F interval. Therefore, the P-F interval isn't long enough to accommodate the inspection, so it is not long enough to manage the consequences of failure.

Scenario 2: Assume that the asset with the V-belt is shut down every 3 months. In this scenario, the item can be monitored at intervals less than the P-F interval. Additionally, in this example, the working group identifies that the V-belts are kept in stock and can be changed before the asset is scheduled to be started again. Therefore, in Scenario 2, the net P-F interval of 3 months is long enough to *manage the consequences of failure.*

Using the Decision Diagram to Assign On-Condition Tasks

Figure 9.15 illustrates the evident-safety On-Condition block from the RCM Decision Diagram.

Just like Restoration and Replacement task blocks, the text in the first set of questions within the On-Condition block is normal font. However, the text in the second set of questions is italicized. As illustrated in Figure 9.16, the first set of questions establishes if a task is *technically appropriate.* The second set of questions establishes if the task is *worth doing.*

The facilitator asks the questions and the working group provides the answers. If the answer to any one of the questions within the block is *No*, then that task is not appropriate to manage the Failure Mode. The facilitator leads the working group to the next block. However, if the answer to all of the questions is *Yes,* then that task is assigned to manage the Failure Mode. The following example illustrates these points.

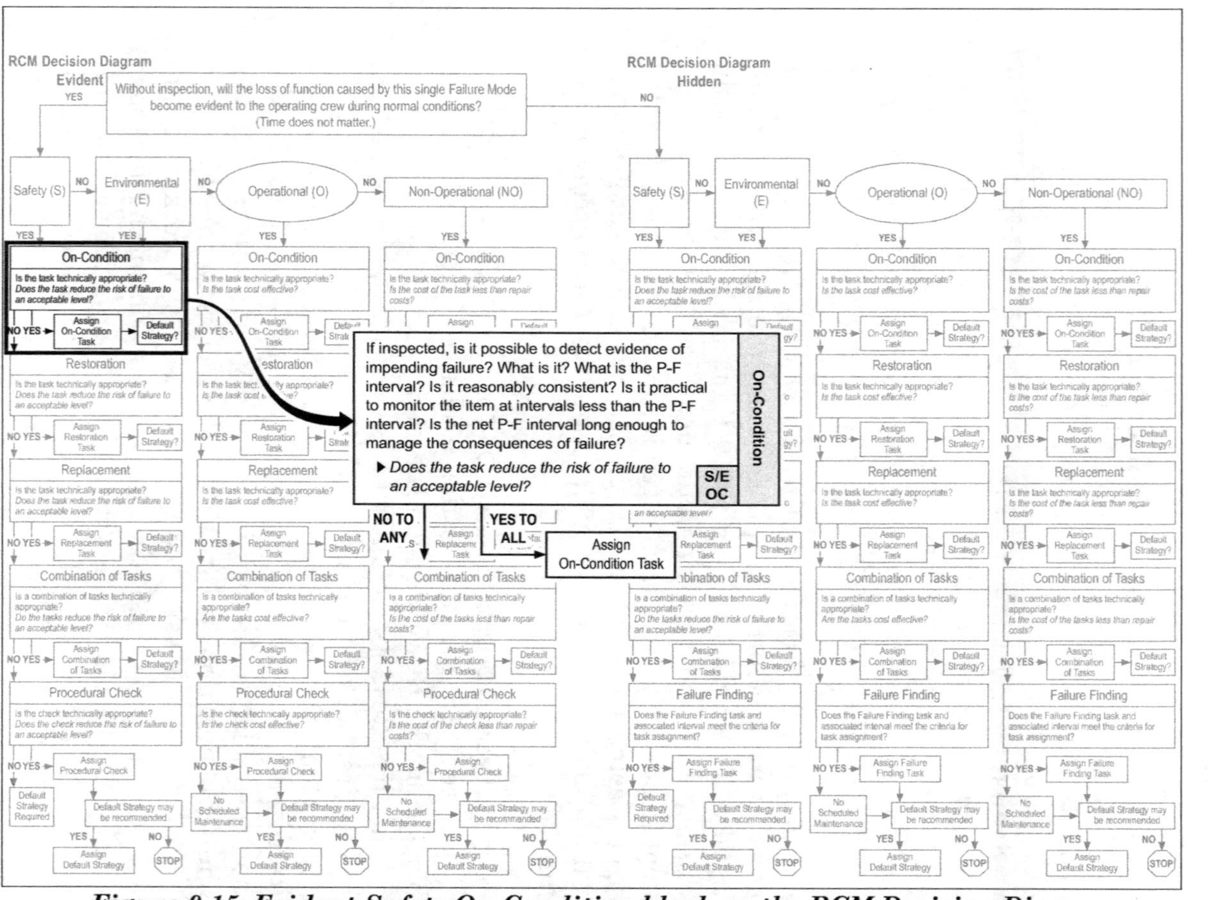

Figure 9.15 Evident-Safety On-Condition block on the RCM Decision Diagram

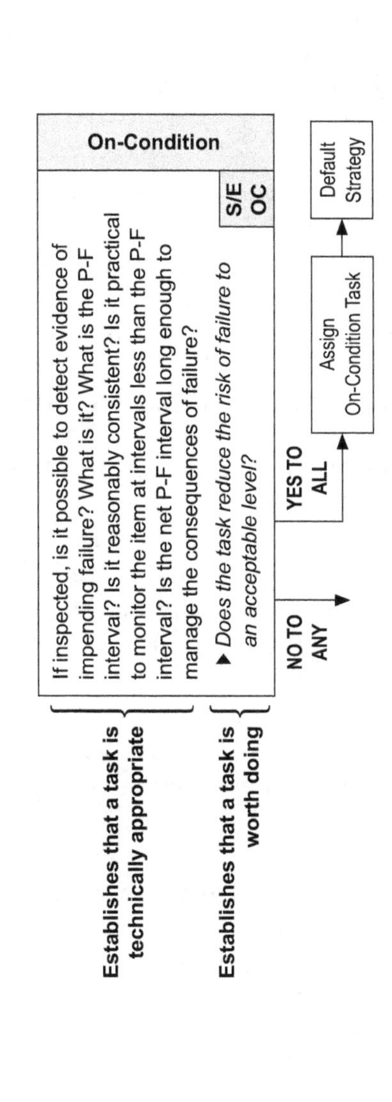

Figure 9.16 Establishing that an On-Condition task is technically appropriate and worth doing for evident Failure Modes with safety/environmental consequences.

Example

Failure Mode: *Tow tractor front brake pads wear due to normal use.*

In this example, the working group identified that the Failure Mode is evident and that it has safety consequences. From there, the facilitator led the working group to the On-Condition block where the following questions were asked and answered in order to determine if an On-Condition task is *technically appropriate* and *worth doing.*

If inspected, is it possible to detect evidence of impending failure? *Yes*

What is it? Brake pads worn to only 1/16″ remaining

What is the P-F interval? 1 year (The working group identified that what they consider "failure" is the brake drums being damaged.)

Is the P-F interval reasonably consistent? *Yes*

Note about the P-F interval and consistency Oftentimes P-F intervals may vary, especially when an organization has several of the same asset. Just because P-F intervals vary doesn't necessarily mean the answer to the question *"Is the P-F interval reasonably consistent?"* is *No.* Suppose individual working group members identify a P-F interval as 1 year, 14 months, and 18 months. These three answers are inconsistent. However, the facilitator can query the working group and ask, *"Is it consistently at least as long as 1 year?"* If the answer is *Yes*, then the P-F interval of 1 year has been established as being consistent.

Is it practical to monitor the item at intervals less than the P-F interval? *Yes*

Is the net P-F interval long enough to manage the consequences of failure? *Yes*

At this point, the working group has answered affirmatively to the first set of questions, thereby determining that it is technically appropriate to measure the brake pads. They decide to halve the P-F interval and inspect the brake pads every 6 months. Now the working group must determine if the task is

RCM Decision Worksheet

Failure Mode	E	ES	EE	EO	ENO	HS	HE	HO	HNO	OC	Rst	Rpl	C	PC	FF	DS	Task	Initial Interval	Default Strategy
5 A 1	Y	Y								Y							Measure the brake pads. Replace the brake pads when there is less than 1/16" remaining.	6 months	

Figure 9.17: Documenting an On-Condition task on the RCM Decision Worksheet

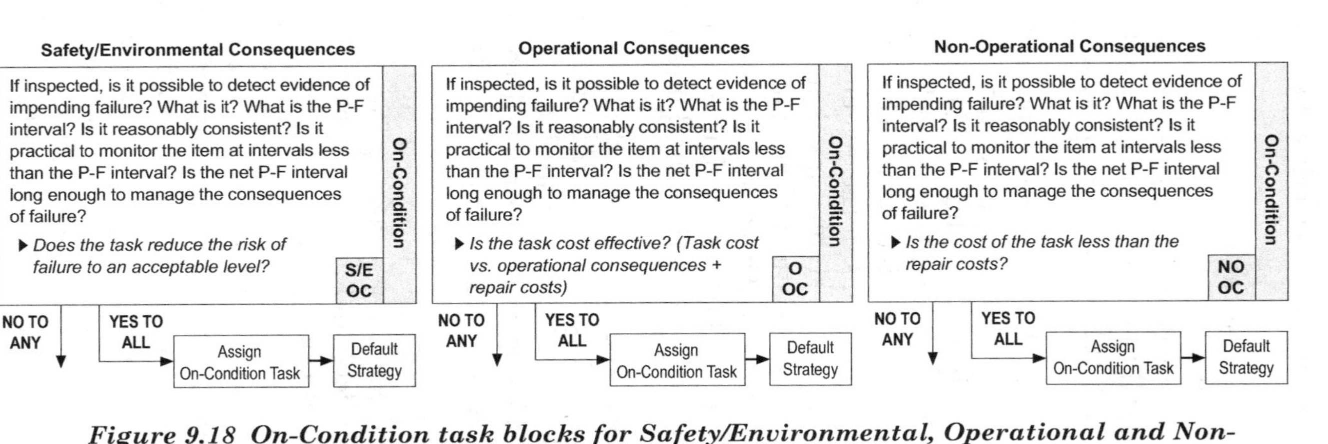

Figure 9.18 On-Condition task blocks for Safety/Environmental, Operational and Non-Operational Consequences

worth doing by answering the last question.

Does the task reduce the risk of failure to an acceptable level? *Yes*

The working group determined that safety will be maintained if the brake pads are monitored every 6 months. The net P-F interval, 6 months, is long enough to replace the brake pads.

Figure 9.17 illustrates how to document this example on the RCM Decision Worksheet. The Failure Mode is evident so *Y* is placed in the E column. The working group determined that failure of the brakes would have safety consequences so *Y* is placed in the ES column. The working group answered *Yes* to all the questions in the evident-safety block depicted in Figure 9.16, thereby establishing that the On-Condition task is technically appropriate and worth doing. Therefore, *Y* is placed in the On-Condition (OC) column as shown in Figure 9.17. The task is written in the task column and the initial interval of 6 months is documented in the initial interval column.

Differences amongst On-Condition Task Blocks for Safety/Environmental, Operational, and Non-Operational Consequences

Figure 9.18 illustrates the On-Condition blocks from the RCM Decision Diagram for all consequences. There are few differences amongst the blocks. The differences are discussed below.

Worth Doing: Risk versus Cost As noted earlier in this chapter and as depicted in Figure 9.19, when dealing with safety or environmental consequences, the way to measure whether a task is worth doing is by assessing risk. That is, the task has to reduce the risk of failure to an acceptable level. On the other hand, when dealing with operational and non-operational consequences, worth doing is determined by assessing cost. For operational consequences, if it costs less to do the task versus what it would cost to bear the operational consequences plus repair costs, then the task is cost effective. For non-operational consequences, the task is cost effective if the cost of the task is less than the repair costs.

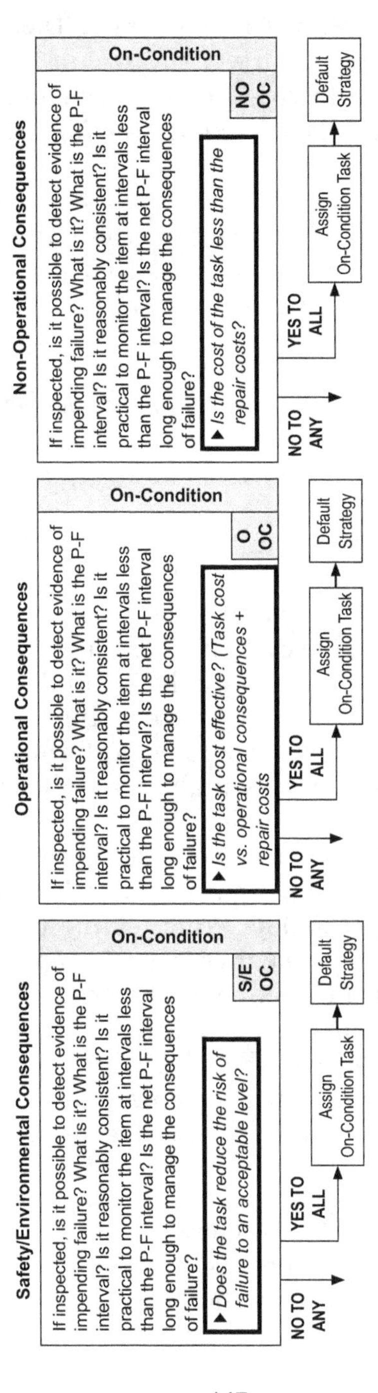

Figure 9.19 Risk versus cost when determining if an On-Condition task is worth doing

Determining if a Proactive Maintenance Task is Cost Effective The following example illustrates how a working group can determine if a proactive task is cost effective. Consider the following example:

Failure Mode: *Feedwater pump bearing wears due to normal use.*

The working group concludes that the Failure Mode is evident-operational and they determine that performing the On-Condition task is technically appropriate.

On-Condition task: *Every 3 months, perform vibration analysis on feedwater pump bearing in accordance with SB-54008. Replace the bearing, as required.*

Now the working group must determine if the task is cost effective. Oftentimes, determining if a task is cost effective is largely an intuitive process and a cost analysis isn't necessary to make the decision. However, if it is, the facilitator leads the working group through a cost analysis similar to the one depicted in Figure 9.20 for the feedwater pump bearing.

Note that a cost analysis should be performed over a significant period of time. In the cost analysis detailed in Figure 9.20, if vibration analysis is performed over a 12-year period, it costs the organization $14,400. However, if the bearing is allowed to run to failure, the cost of the unanticipated bearing failure over the 12-year period is $60,000. Whether the proactive maintenance is done or not, the bearings must be replaced, so the cost of doing so cancels out. Notice that when calculating the cost of not doing any maintenance, the MTBF of the bearing is used. It is appropriate to use MTBF when determining if a task is worth doing because the MTBF tell us how often, on average, the bearing will fail. Clearly in this case, it is cost effective to do the maintenance.

Additional Issues Regarding the P-F Curve

The P-F interval examples used above are straightforward and explain the concepts embodied within the P-F curve. However, assigning variables on the P-F curve can be more intri-

Failure Mode: Feedwater pump bearing wears due to normal use.
MTBF of the bearing: 4 years

Performing Proactive Maintenance

Every 3 months perform vibration analysis on feedwater pump bearing in accordance with SB-544008.

Cost to perform the task: $300

Cost of performing the task over a 12 year period:
(perform task 48 times)($300 per task) =
$14,400 + ~~cost to replace the bearings~~

No Scheduled Maintenance

The feedwater pump bearing is run to failure.

Cost of one unanticipated bearing failure (cost to recycle scrap paper plus production downtime): $20,000

Cost of running the bearing to failure over a 12 year period:
The MTBF of the bearing is 4 years. Therefore, on average, over a 12 year period, the bearing will fail three times.

(3 unanticipated bearing failures)($20,000) =
$60,000 + ~~cost to replace the bearings~~

Figure 9.20 Cost analysis for feedwater pump bearing

cate. The following discussion sets forth examples that illustrate more complex uses of the P-F curve that may arise during an analysis.

Example 1: Gauge Accuracy and Tolerance

Function: *To display the quantity of fuel in the tank within +/–10% of actual.*

Functional Failure: *Displays the quantity of fuel in the tank within more than +/–10% of actual.*

Failure Mode: *Fuel level indicator drifts out of adjustment due to normal use.*

The facilitator leads the working group in identifying variables for the P-F curve, as depicted in Figure 9.21. First, the working group identifies the accuracy of the gauge. This measure represents how accurate the gauge will be when received from the manufacturer. In this case, the best case is that the gauge will arrive out of adjustment +/–3%. The tolerance repre-

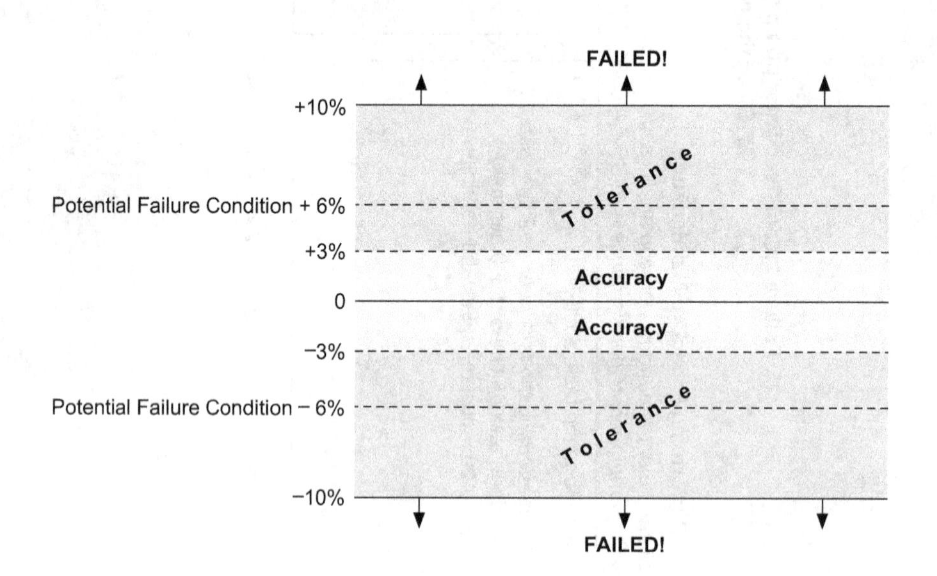

Figure 9.21 Example of a fuel quantity gauge: Accuracy versus Tolerance and the P-F Interval

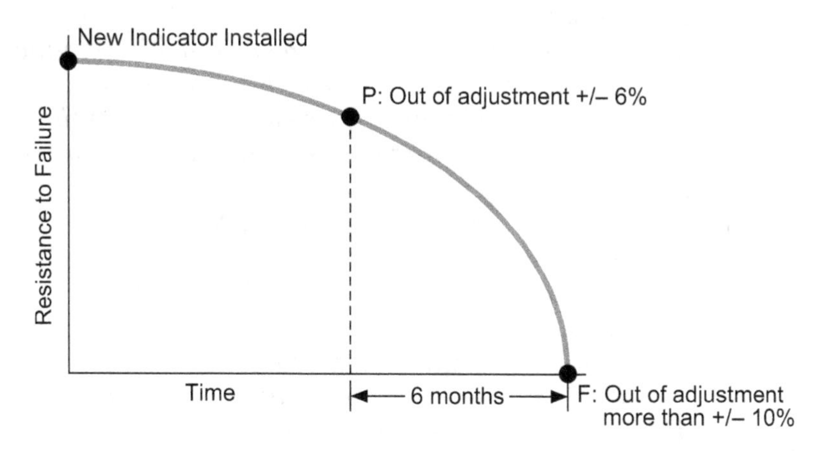

Figure 9.22 Example of a fuel quantity gauge: Accuracy versus Tolerance and the P-F Interval

sents how far out of adjustment the organization is willing to accept. The organization is willing for the gauge to be out of adjustment +/–10%. Anything beyond +/–10% is considered failed by the organization.

The facilitator leads the working group in identifying the variables on the P-F curve for the fuel quantity gauge, as depicted in Figure 9.22.

The working group identifies that the potential Failure condition is +/–6%. That is, +/–6% is still within acceptable range for the organization, but it signifies that failure (+/–10%) is impending. The working group members identify failure (F) as the gauge being out of adjustment more than +/–10%. Finally, they determine that the P-F interval is 6 months. With this information identified, the working group can answer the questions posed within the on-condition block of the RCM Decision Diagram depicted in Figure 9.19.

Example 2: Identifying "Failure" on the P-F Curve As illustrated in the following example, failure as identified on the P-F Curve is not always the same as the Functional Failure identified on the RCM Information Worksheet.

Engine preservation example

Function: *To prevent corrosion on the gas turbine engine while stored inside the engine shipping container.*

Functional Failure: *Unable to prevent corrosion on the gas turbine engine while stored inside the engine shipping container.*

Failure Mode: *Any engine shipping container seal deteriorates due to normal use.*

Excerpt from Failure Effect: *Humidity is introduced inside the shipping container. Eventually, the desiccant is compromised and the humidity indicator on the exterior of the shipping container indicates saturation (40% humidity). If the saturation indication goes unnoticed, humidity inside the shipping container promotes corrosion.*

The facilitator leads the working group in identifying the variables on the P-F Curve, as depicted in Figure 9.23.

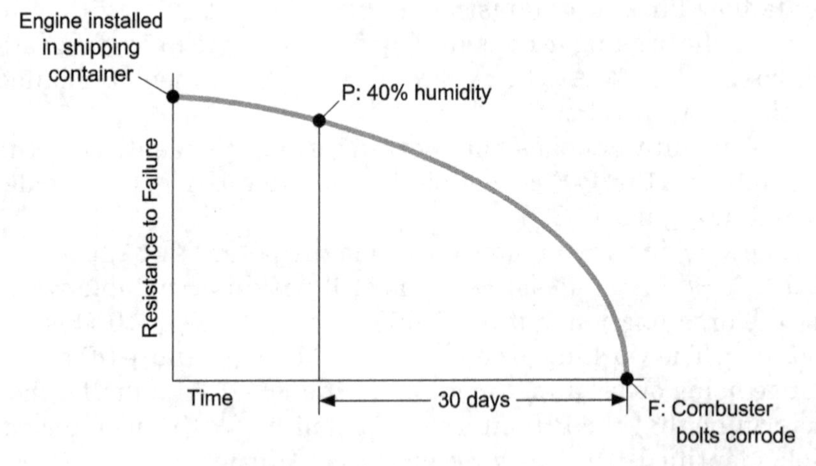

Figure 9.23 Engine preservation P-F Curve Example

The working group identifies that the potential failure condition (P) is 40% humidity inside the container. In discussing what constitutes failure, the group identifies that failure would be the combustor bolts corroding—the combustor bolts would be the first engine items to corrode. If increased humidity can be detected (by visually inspecting the humidity indicator) before the combustor bolts corrode, then no other engine components will be damaged due to corrosion.

Next, the working group discusses the P-F interval. Discussion is given to differing operating environments (e.g., shipping container stored inside versus outside, where the shipping container is exposed to the elements). However, the working group determines that the P-F interval is the same for all operating environments. Once the internal humidity reaches 40%, the P-F interval is the same, regardless of the external environment. The only interval that is longer because of the operating environment is the period of time from when an engine is installed in the shipping container to the point that the humidity reaches 40%. That interval doesn't have anything to do with the P-F interval. When determining P-F intervals, working group members should consider only the time from when the potential failure condition is detectable to the time the failure (as defined by the working group) occurs.

Example 3: Several Potential Failure Conditions Failure Modes can often include several potential failure conditions. The facilitator leads the working group in identifying the possible potential failure conditions and the associated P-F intervals. The work-

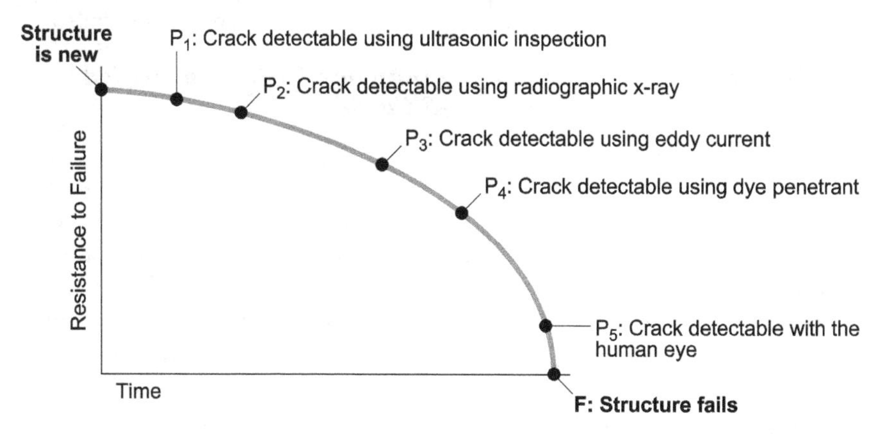

Figure 9.24 Example of several Potential Failure Conditions for one Failure Mode

ing group members then use the questions posed in the On-Condition block on the RCM Decision Diagram to determine which potential failure condition (if any) will be monitored. Consider the following example for the Failure Mode *Structure fatigues due to normal use.* Potential failure conditions for the detection of cracks are depicted in Figure 9.24.

Example 4: More than One Failure Oftentimes, a potential failure condition can include more than one failure, as identified by the working group. Consider the following example for the Failure Mode *XYZ component corrodes due to normal use,* as depicted in Figure 9.25.

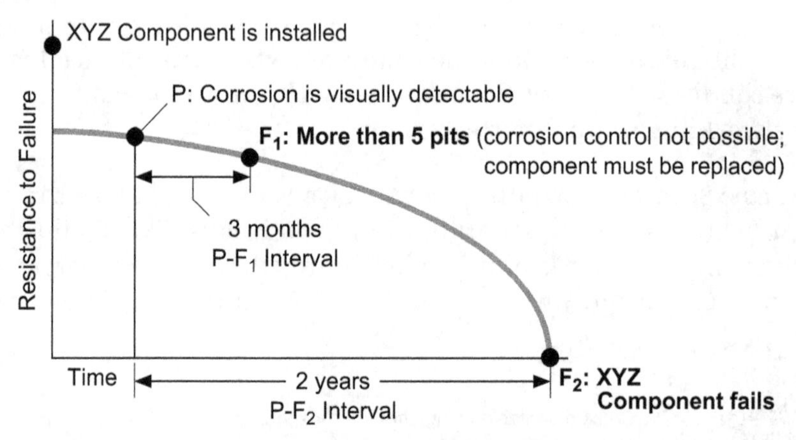

Figure 9.25 Example of more than one Failure identified on the P-F Curve

In this example, F_1 is *more than 5 pits* have developed. If corrosion is allowed to get to the point that more than 5 pits develop, the component will have to be replaced. Having to replace the component is considered a failure to the organization because replacing it is much more expensive and time consuming than simply performing corrosion control. F_2 is *component fails* which has a much longer P-F interval associated with it. If F_1 is used, the component will have to be inspected quite frequently, but the repair (corrosion control) is uncomplicated and quick. If F_2 is used, the component can be inspected much less frequently. However, by the time the corrosion is detected, the component must be replaced—this replacement takes considerable downtime plus the cost of the part itself. During the RCM analysis, the working group should consider each scenario and make the more cost-effective decision for the organization.

Example 5: Safety Factor and On-Condition Task Intervals for Failure Modes with Safety or Environmental Consequences When scheduled Restoration and Replacement task intervals for Failure Modes with safety or environmental consequences were discussed in Section 9.3, the point was made that the useful life must be identified so that no failures will occur before reaching the end of the useful life. However, it was also noted there is no guarantee of zero failures. A safe-life limit is typically established such that no failures are *expected to occur* before the specified life limit. A safety factor may apply to On-Condition tasks as well. For example, if a P-F interval is identified as 1 year, a general rule of thumb is to perform the On-Condition inspection at half the P-F interval—every 6 months. However, in this case, an organization may decide to do the task even more frequently, for example, every 2 months.

Starting an On-Condition Maintenance Program (Commonly Referred to as a Condition-Based Maintenance Program or CBM)

It is common today to hear that organizations are working toward incorporating more sophisticated monitoring devices in an attempt to reduce the amount of scheduled maintenance that is performed. Organizations can benefit by On-Condition maintenance, but On-Condition maintenance must be established the right way.

To properly implement a CBM program, organizations must first identify the Failure Modes they wish to manage using On-Condition maintenance.

Physical assets are managed at the Failure Mode level. To properly implement a CBM program, organizations must first identify the Failure Modes they wish to manage using On-Condition maintenance, applying the process outlined in Section 9.4. However, in many cases, a later step in establishing a

CBM program is accomplished first—procuring the technology. This may result in data being captured that equipment custodians don't quite know what to do with. On some occasions, sophisticated monitoring devices *could* be used, but it may be more cost effective to perform a visual inspection or to monitor a gauge. In order to implement a CBM program properly, Failure Modes should be identified first. Then the questions in the On-Condition block on the RCM Decision Diagram must be answered for each Failure Mode identified. These steps will ensure that the safest and most cost-effective solutions are formulated.

Important Points Regarding On-Condition Tasks

On-Condition maintenance tasks can be very powerful failure management strategies. Remember these important points about On-Condition maintenance.

- On-Condition maintenance allows a potential failure condition to be detected in enough time to manage the consequences of failure. The consequences of failure can vary. For example, a hydraulic system warning light is heeded that allows enough time for the pilot to *land the aircraft safely*. A specific vibration level is detected in enough time to order the bearing and schedule the maintenance so that *operations are not adversely affected*. A low oil level is discovered so that the reservoir can be replenished *before the engine is damaged*.

- The P-F interval dictates how often an On-Condition task is performed. In other words, what matters is *how quickly the failure occurs once the potential failure condition is detectable*. Frequency and criticality of failure *do not* dictate how often an On-Condition task is performed. MTBF is *not* used to set On-Condition task intervals. Furthermore, the length of time between when a component is installed and when the potential failure condition becomes detectable *has nothing to do* with how often an On-Condition task is performed.

- Identifying P-F intervals is often a difficult task. Rarely are there databases full of information that can be used to derive P-F intervals. For less sophisticated potential failure conditions—such as audible noise, wear, and visual evidence of corrosion—working group members do a remarkable job at identifying P-F intervals. Working group members are intimate with both the equipment and the operating environment, and they understand how failure occurs. However, when it comes to sophisticated monitoring devices, quantifying the potential failure condition and the associated P-F interval may require considerable investment; therefore, it must be beneficial in order to engage in the research.

On-Condition maintenance is a powerful failure management strategy because it allows impending failure to be identified before the failure occurs. Proactive action can then be taken in enough time to manage the consequences of failure. In other words, maintenance is performed on the *equipment custodian's* terms—not *the equipment's* terms.

9.5 Combination of Tasks

The combination of tasks is represented on the RCM Decision Diagram in the fourth row, as outlined in Figure 9.26.

Combination of tasks means that for one Failure Mode, more than one task is assigned. These tasks are a combination of On-Condition, Restoration, or Replacement tasks. Combination of tasks is rarely used.

An example of a Failure Mode that requires a combination of tasks is detailed below.

Example

Failure Mode: *Suspended particulates in supply water clog the water side of any 1 of 4 intercoolers due to normal use*

In this example, the working group identified a combination of tasks:

- **Restoration task:** *Backflush each interstage cooler weekly.*

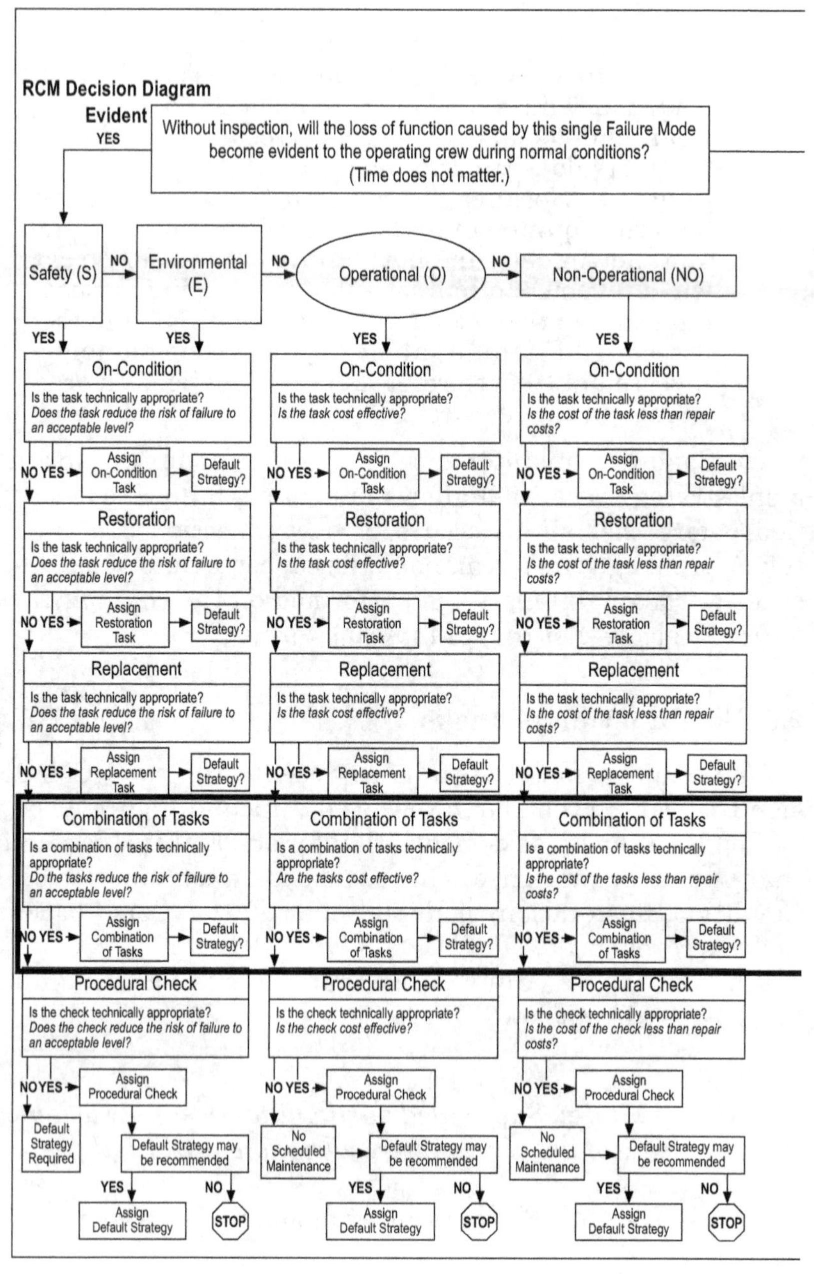

Figure 9.26 Combination of Tasks represented on the RCM Decision Diagram

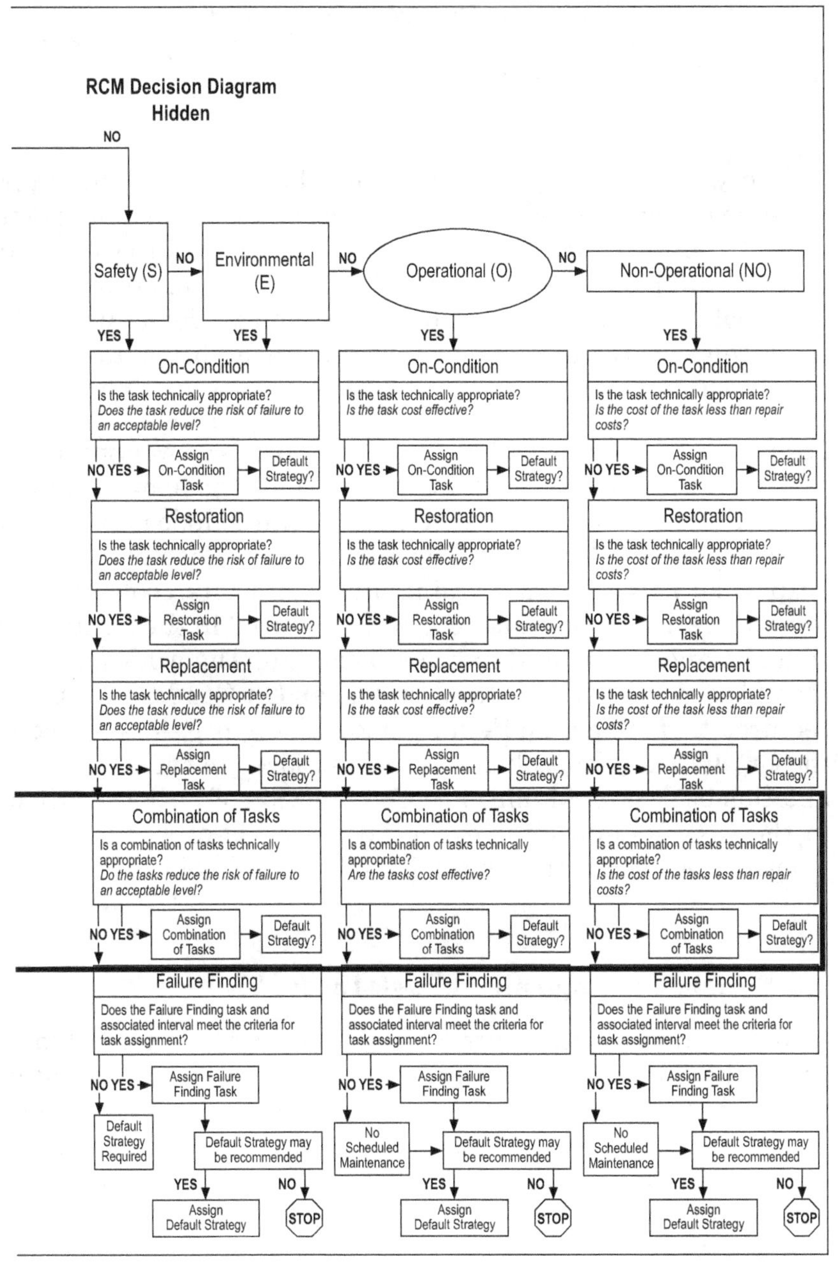

Figure 9.26 Continued

- **On-Condition task:** *Record the fourth inter-stage temperature every three months. If the temperature is 110°F or higher, clean the intercooler.*

The working group identified that backflushing the intercooler every week is effective, but it doesn't remove all particulates. In spite of backflushing, eventually particulates build up and the intercooler temperature increases. By monitoring the intercooler temperature, the organization can detect when the intercooler is required to be cleaned, in addition to weekly backflushing.

Documenting Combination of Tasks on the RCM Decision Worksheet

Figure 9.27 illustrates how to document the above example on the RCM Decision Worksheet. The Failure Mode is evident so *Y* is placed in the E column. The working group determined that suspended particulates clogging any 1 of 4 intercoolers would have operational consequences, so *N* is placed in the ES and EE columns and *Y* is placed in the EO column. The working group identified that both a Restoration and an On-Condition task are required to manage the Failure Mode. Therefore, *N* is placed in the OC, Rst, and Rpl columns and *Y* is placed in the C (for Combination of Tasks) column. The Combination of Tasks is written in the task column and the initial intervals of weekly and 3 months are documented in the initial interval column.

9.6 Synchronizing Initial Task Intervals

When entering task intervals on the RCM Decision Worksheet, the *initial interval* should be recorded. The initial interval is whatever interval is decided upon by the working group as a result of the P-F interval or useful life. A completed analysis typically yields wide-ranging intervals such as: prior to use; 50, 60, 65, 100, and 110 operating hours; 30 days; 6 months; 1 year; etc.

As one of the last steps in the analysis, the facilitator leads the working group in synchronizing the task intervals into

RCM Decision Worksheet

Failure Mode			E	ES	EE	EO	ENO	HS	HE	HO	HNO	OC	Rst	Rpl	C	PC	FF	DS	Task	Initial Interval	Default Strategy
1	B	3	Y	N	N	Y						N	N	N	Y			N	Backflush each interstage cooler Record the fourth interstage temperature every three months. If the temperature is 110°F or higher, clean the intercooler	Weekly 3 months	

Figure 9.27 Documenting a Combination of Tasks on the RCM Decision Worksheet

maintenance packages that make sense. For example, the working group may synchronize three task intervals of 50, 60, and 65 operating hours to 50 operating hours so all three tasks are done at once.

Working group members naturally want to synchronize as they go because having three maintenance packages of 50, 60, and 65 operating hours doesn't make sense. However, it is very important to make sure initial intervals are recorded because the record preserves the audit trail. Ensuring that these initial intervals are recorded on the RCM Decision Worksheet is another reason why a trained facilitator must lead the working group.

Concluding Thought about the RCM Decision Diagram and Proactive Maintenance

The facilitator leads the working group through the RCM Decision Diagram, in the order it is laid out, to analyze whether or not a proactive maintenance task is technically appropriate and worth doing. The RCM Decision Diagram implies that On-Condition tasks are favored before Restoration and Replacement tasks. Generally speaking, this is true because On-Condition tasks allow a component to stay in service as long as possible, thus realizing its maximum useful life. However, a facilitator is trained to lead the working group through the process. At times, a lower order task (such as Replacement) is clearly the way to go for a particular Failure Mode, and the facilitator leads working group members to that conclusion. This is yet another example of why it is so important to have a trained facilitator lead the team.

Summary

There are two types of proactive maintenance: preventive maintenance and On-Condition maintenance. Preventive maintenance includes only scheduled Restoration and scheduled Replacement tasks. In the context of RCM, in order to schedule a proactive maintenance task, two criteria must be satisfied. The proactive task must be 1) technically appropriate and 2) worth doing. The RCM Decision Diagram provides tools for determining if the two criteria are satisfied.

A scheduled Restoration task reworks or restores an item's failure resistance to an acceptable level. A scheduled Replacement task replaces an item. Both tasks are performed at fixed intervals without considering the item's condition at the time of the task. In the context of RCM, scheduled Restoration and Replacement tasks are considered preventive maintenance because action is taken before failure is allowed to occur, thereby preventing failure. One of the most important elements that makes a scheduled Restoration or Replacement task technically appropriate is if there is a useful life.

Most Failure Modes provide evidence that failure is in the process of occurring. In the context of RCM, evidence of impending failure such as vibration, heat, noise, wear, warning lights, and gauge indications are called potential failure conditions. An On-Condition task is performed at a defined interval to detect the evidence of impending failure. Therefore, maintenance is performed only on the evidence of need and the organization gets the maximum useful life out of a component. On-Condition task intervals are determined based upon the P-F interval, which is the time from when the potential failure condition is detectable to the point that failure occurs.

The facilitator leads the working group through the RCM Decision Diagram in order to determine if a proactive maintenance task is technically appropriate and worth doing. All decisions are recorded on the RCM Decision Worksheet.

Ten

Default Strategies

Assigning Default Strategies is the seventh and last step in the RCM process. This chapter discusses the formulation of:

- Procedural Checks
- Failure Finding Tasks
- No scheduled maintenance
- Other Default Strategies such as engineering redesigns, operating procedure additions, supply changes, technical publication revisions, and training recommendations
- Notes regarding Default Strategies

10.1 Procedural Checks

The Procedural Check block was added to the RCM Decision Diagram because oftentimes a check is required to manage a Failure Mode that does not fall under the realm of traditional RCM task definition.

Procedural Checks are unique to the RCM Decision Diagram included in *The RCM Solution*. The Procedural Check block was added to the RCM Decision Diagram because oftentimes a check is required to manage a Failure Mode that does not fall under the realm of traditional RCM task definition. Procedural Checks are found only on the evident side of the RCM Decision Dia-

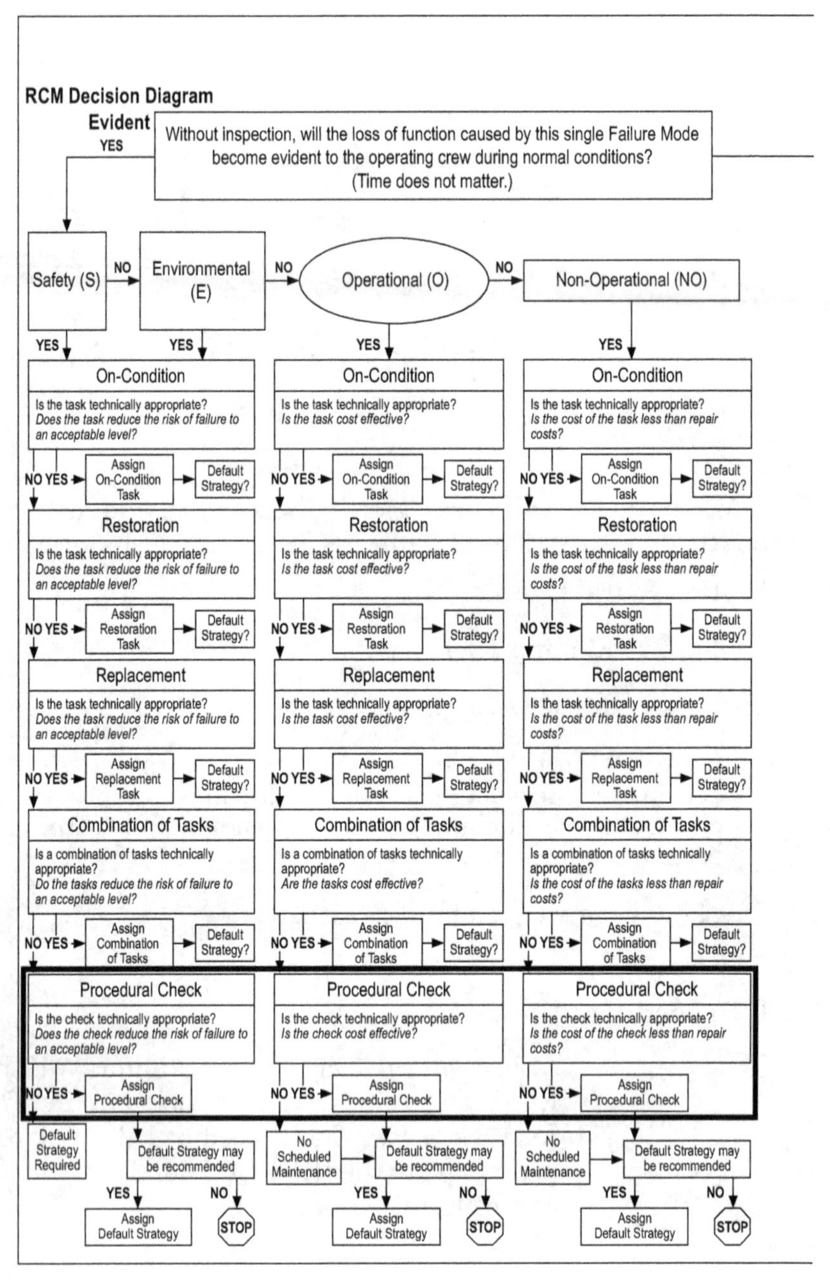

Figure 10.1 Procedural Tasks represented on the RCM Decision Diagram

gram. A Procedural Check is the last type of task that is considered for evident Failure Modes, as depicted in Figure 10.1.

In the context of RCM, a Procedural Check is considered a Default Strategy because it isn't a proactive task; it doesn't meet the criteria for an On-Condition, Restoration, or a Replacement task. A Procedural Check is performed to check for a failure that may have already occurred because it would be unacceptable to operate, or continue to operate, with the failure condition.

Procedural Check

A task performed at a specified interval to check for a failure that may have already occurred because it would be unacceptable to operate or continue operating with the failure condition.

The addition of Procedural Checks to the RCM Decision Diagram allows equipment custodians to identify tasks that may be included in checklists such as pre-operational checks, post-operational checks, and round sheets. The capability of doing so does not exist in traditional RCM Decision Diagrams.

Criteria for Assigning a Procedural Check

In the context of RCM, just like proactive tasks, Procedural Checks must be technically appropriate and worth doing in order to schedule the check. The questions embodied in the Decision Diagram within the Procedural Check block are answered in order to determine if a Procedural Check is technically appropriate and worth doing.

Using the Decision Diagram to Assign Procedural Checks

Figure 10.2 illustrates the safety-environmental Procedural Check block from the RCM Decision Diagram for all consequences. Notice that the text in the first set of questions within each block is normal font. However, the text in the second set of questions is italicized. As illustrated in Figure 10.3, the first set of questions establishes if a task is *technically ap-*

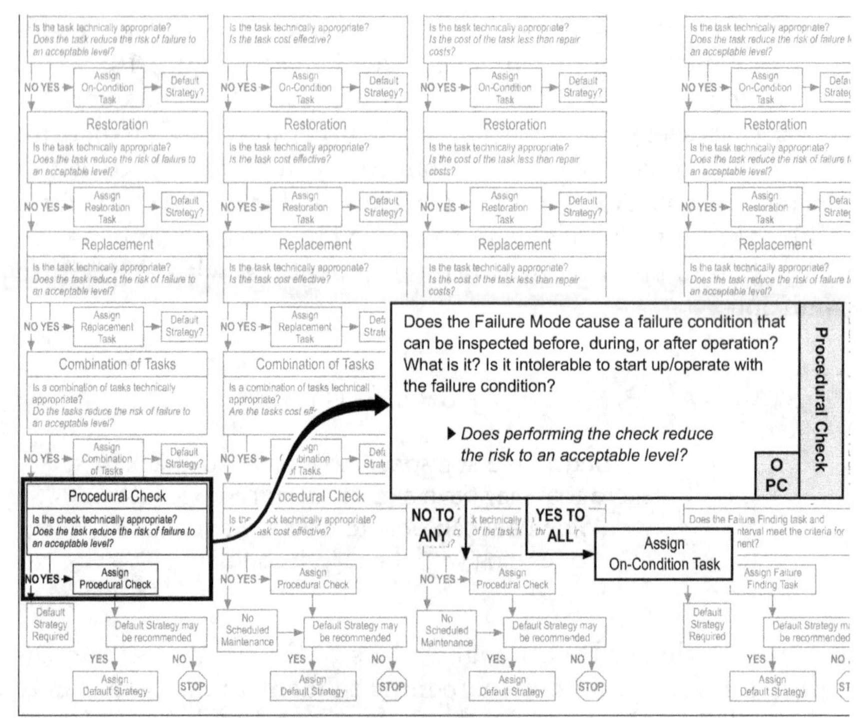

Figure 10.2 Safety-Environmental Procedural Check block on the RCM Decision Diagram

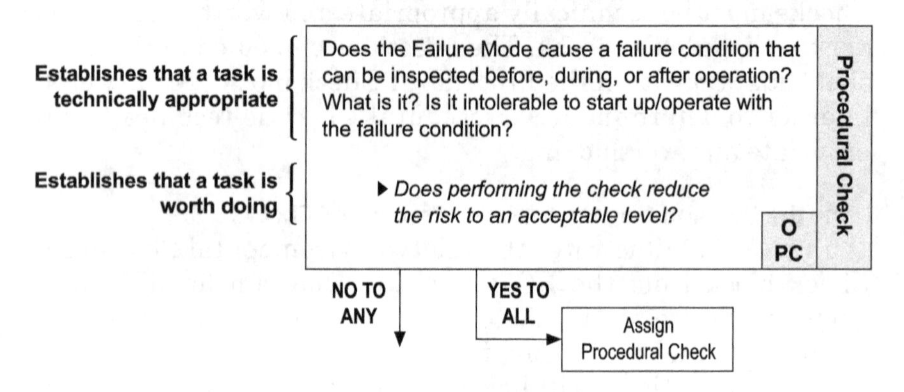

Figure 10.3 Establishing that a Procedural Check is technically appropriate and worth doing

propriate. The second set of questions establishes if the task is *worth doing*.

The facilitator asks the questions and the working group provides the answers. If the answer to any one of the questions within the block is *No*, then the Procedural Check is not appropriate to manage the Failure Mode. The facilitator leads the working group to the next block. However, if the answer to all of the questions is *Yes*, then that Procedural Check is assigned to manage the Failure Mode. The following example illustrates these points.

Example

Failure Mode: *Aircraft fuselage damaged by support equipment*

In this example, the working group identified that the Failure Mode is evident and that it has safety consequences. From there, the facilitator led the working group through the evident-environmental column on the RCM Decision Diagram. The working group answers *No* to the first question in the On-Condition block: *If inspected is it possible to detect evidence of impending failure?* The answer is *No* because there is no evidence that the failure is in the process of occurring—it has already occurred. Therefore, the facilitator leads the working group to the Restoration and Replacement blocks.

Similarly, the working group answers *No* to the first question in each block: *Is there an identifiable wearout age?* There certainly is no wearout age. The support equipment striking the aircraft is a random event. If an On-Condition, Restoration, or Replacement task isn't technically appropriate, then a Combination of tasks is not. Therefore, the facilitator leads the working group to the Procedural Check block where the following questions are asked and answered.

Does the Failure Mode cause a failure condition that can be inspected before, during, or after operation? *Yes, prior to flight.*

What is it? *Physical damage done to the aircraft*

Is it intolerable to start up/operate with the failure condition? *Yes, the pilot should not take off with part of the aircraft damaged.*

At this point, the working group has answered affirmatively to the first set of questions, thereby determining that it is technically appropriate to inspect the aircraft for damage prior to flight. Now the working group must determine if the task is worth doing by answering the last question.

Does performing the check reduce the risk to a tolerable level? *Yes*. The working group identifies that inspecting the aircraft for damage prior to flight reduces the risk to an acceptable level.

Because all of the questions have been answered affirmatively, the Procedural check is assigned.

Figure 10.4 illustrates how to document the above example on the RCM Decision Worksheet. The Failure Mode is evident, so a *Y* is placed in the E column. The working group determined that damage to the aircraft could adversely affect safe flight, so *Y* is placed in the ES block. The working group answered *No* to On-Condition, Restoration, Replacement tasks and Combination of tasks, so *N* is placed in the OC, RST, RPL, and C columns, respectively. *Yes* was answered to all the questions in the evident-safety Procedural Check block depicted in Figure 10.3 establishing that the Procedural Check is technically appropriate and worth doing. Therefore, *Y* is placed in the PC column. The task is written in the Default Strategy column and the initial interval of "prior to flight" is documented in the initial interval column.

Procedural Checks are effective for evident Failure Modes that cannot be predicted or prevented such as:

- Fastener on access panel not latched properly
- Fuel line leaks due to normal use
- Tow tractor tire is penetrated by FOD
- Hydraulic fill cap not replaced properly

All of these Failure Modes cause a failure condition as soon as the Failure Mode occurs. However, an organization would most likely consider it intolerable to start up or operate with the failure condition. Procedural Checks are an excellent op-

RCM Decision Worksheet

Failure Mode			E	ES	EE	EO	ENO	HS	HE	HO	HNO	OC	Rst	Rpl	C	PC	FF	DS	Task	Initial Interval	Default Strategy
1	A	4	Y	Y								N	N	N	N	Y		N		Prior to Flight	Inspect the aircraft for damage.

Figure 10.4 Documenting a Procedural Check on the RCM Decision Worksheet

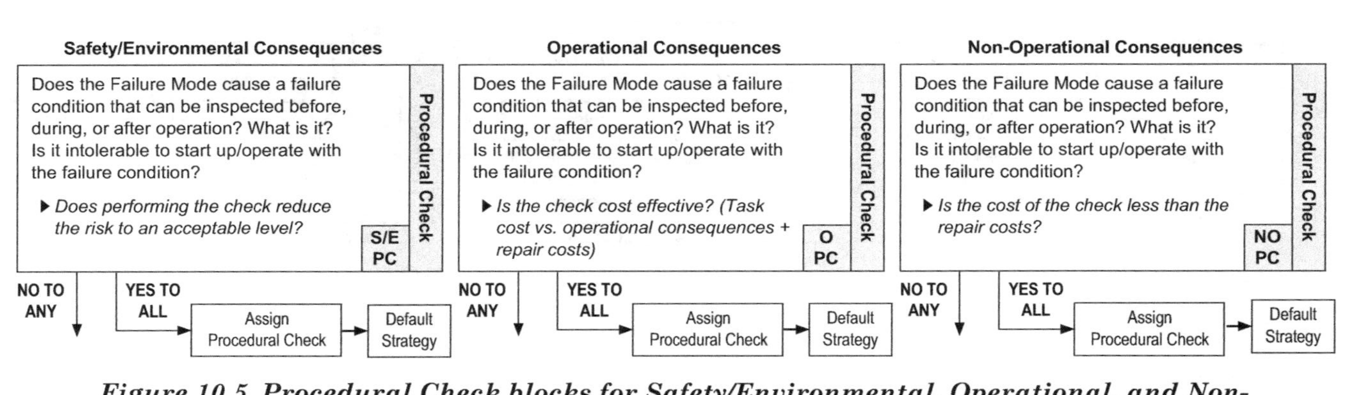

Figure 10.5 Procedural Check blocks for Safety/Environmental, Operational, and Non-Operational Consequences

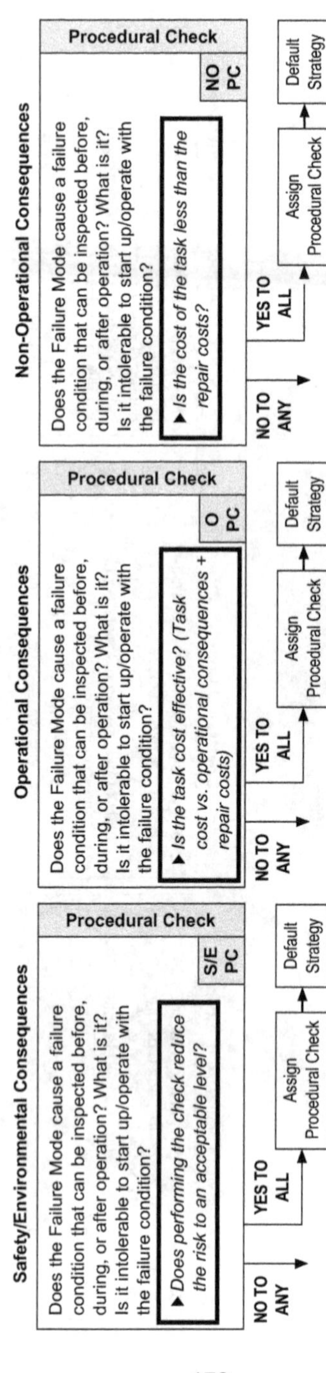

Figure 10.6 Risk versus cost when determining if a Procedural Check is worth doing

portunity to manage human error (e.g., tool left behind) and equipment that is serviced regularly, susceptible to damage, or exposed to traffic.

Differences amongst Procedural Check Blocks for Safety/Environmental, Operational, and Non-Operational Consequences

Figure 10.5 illustrates the Procedural Check blocks from the RCM Decision Diagram for all consequences.

Worth Doing: Risk versus Cost As depicted in Figure 10.6, when dealing with safety or environmental consequences, worth doing is determined by assessing *risk*. That is, the check has to reduce the risk of failure to an acceptable level. On the other hand, when dealing with operational and non-operational consequences, worth doing is determined by assessing *cost*. For operational consequences, if the check costs less to do the task versus what it would cost to bear the operational consequences plus repair costs, then the check is cost effective. For non-operational consequences, the check is cost effective if the cost of the task is less than the repair costs.

10.2 Failure Finding Tasks

Failure Finding tasks are the last type of task that is considered for hidden Failure Modes, as depicted in Figure 10.7.

In the context of RCM, a Failure Finding task is considered a Default Strategy because it isn't a proactive task; it seeks to *find a failure of a protective device* that has already occurred.

Failure Finding Task

A task that checks if a protective device is in a failed state.

As discussed in Chapter 4, Section 4.3, a protective device is a device or system intended to protect the asset, organization, and people *in the event that another failure occurs*.

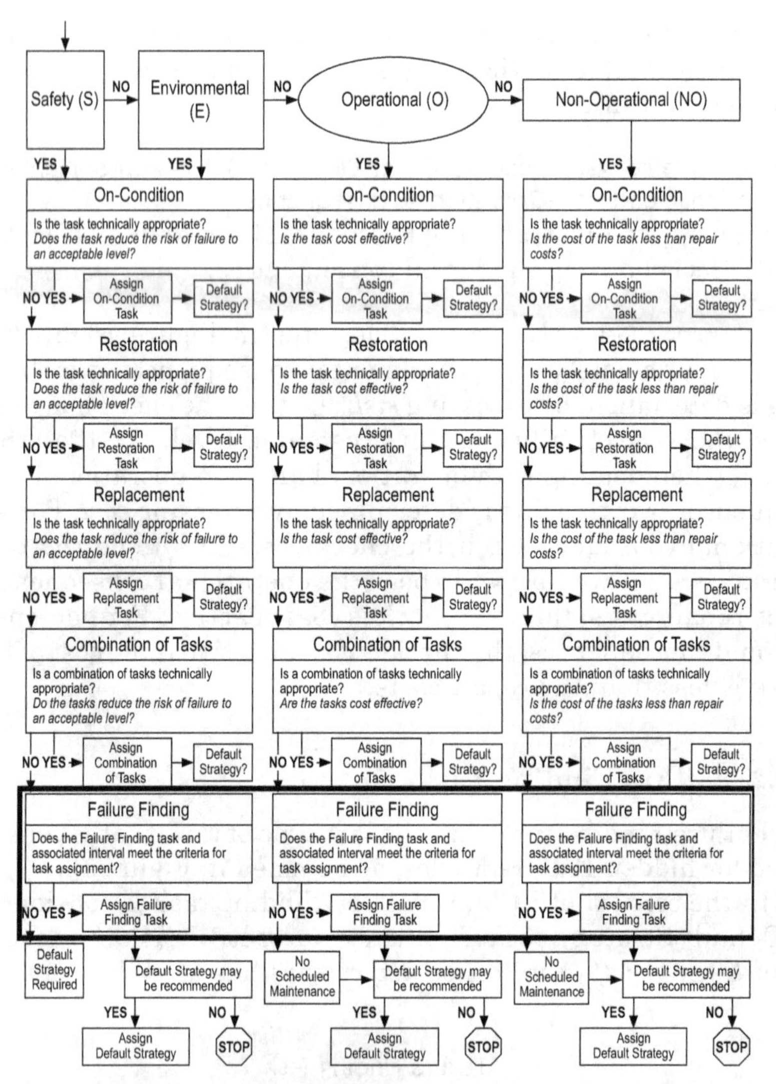

Figure 10.7 Failure Finding represented on the RCM Decision Diagram

Protective Device

A device or system intended to protect people, the asset, and the organization in the event that another failure occurs.

For example, a smoke detector is a protective device. It is required to be capable of sounding an audible alarm *in the event that* there is a fire in the room. Other examples of protective devices are engine chip detector systems, life rafts, emergency exit lighting, emergency stops, fuses, and high temperature cutoff systems.

Purpose of a Failure Finding Task

The purpose of a Failure Finding task is to reduce the risk of a multiple failure to an acceptable level.

The purpose of a Failure Finding task is to reduce the risk of a multiple failure to an acceptable level. For example, if there is a fire in the building, but the smoke detector is operational, the detector sounds an alarm so that building occupants know to evacuate quickly. In the event of a fire and the smoke detector sounds, there is much less chance that building occupants will be injured from the fire. However, if there is a failure of the #2 engine bearing that produces metal particles, but the engine chip detector is in a failed state, the operator is unaware of the impending bearing failure and cannot respond accordingly.

Assigning Failure Finding Tasks

In the context of RCM, in order to assign a Failure Finding task, it must be possible to check the system for failure. Furthermore, the protective device must be checked in its entirety. For example, it is possible to check a smoke detector *in its entirety* by using a spray can of smoke detector test aerosol and simulating smoke at the smoke detector. (Note: pushing and holding the test button typically just tests the alarm portion of the smoke detector.) Conversely, other than ensuring that the correctly-rated fuse is installed and using an Ohmmeter to verify that the fuse element is intact, the act of verifying proper operation of a fuse due to overcurrent destroys the fuse.

Failure Finding task calculations are used to determine Failure Finding task intervals. When calculating intervals, factors such as the following are taken into consideration:

- How often the protective device fails
- How often the protective device is required to work

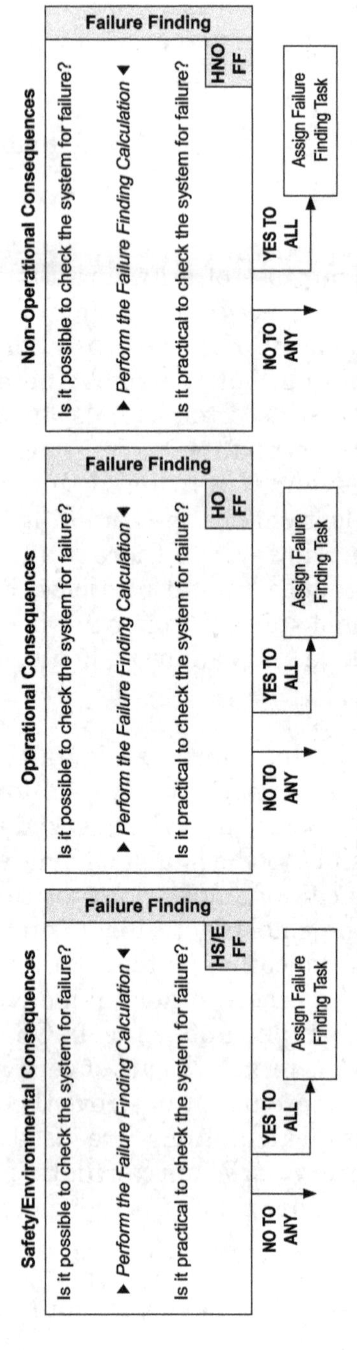

Figure 10.8 Failure Finding blocks for Safety/Environmental, Operational, and Non-Operational Consequences

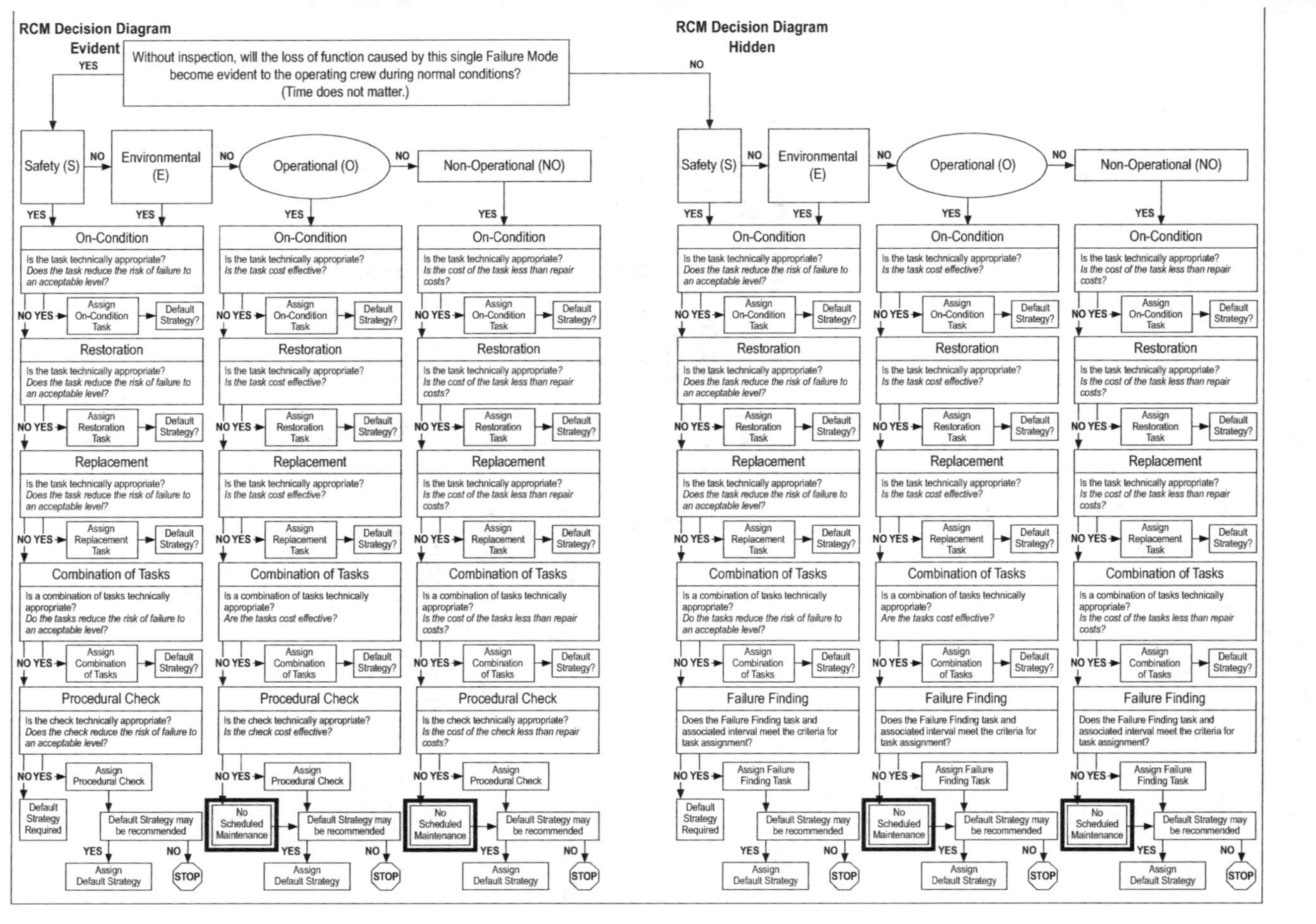

Figure 10.9: No Scheduded Maintenance represented on the RCM Decision Diagram

- How often the organization is willing to accept the multiple failure

The matter of calculating Failure Finding task intervals itself is beyond the scope and purpose of this book. However, calculating Failure Finding task intervals is another reason why it is critical that a trained facilitator lead a working group. Facilitators are trained to employ various formulae with working group members in order to calculate Failure Finding intervals.

Failure Finding Tasks and the Decision Diagram

Figure 10.8 illustrates the Failure Finding blocks from the RCM Decision Diagram for all consequences.

10.3 Synchronizing Initial Task Intervals

As outlined in Chapter 9, Section 9.6 for proactive tasks, when entering task intervals on the RCM Decision Worksheet for Procedural Checks and Failure Finding tasks, the *initial interval* should be recorded in order to preserve the audit trail. In one of the last steps in the analysis, the facilitator leads the working group in synchronizing the task intervals into maintenance packages that make sense.

10.4 No Scheduled Maintenance

No scheduled maintenance appears on the evident and hidden sides of the RCM Decision Diagram under Operational and Non-Operational Consequences, as depicted in Figure 10.9.

No scheduled maintenance to a Failure Mode simply means that maintenance is not the answer to manage the Failure Mode

In the context of RCM, no scheduled maintenance is considered a Default Strategy. Assigning no scheduled maintenance to a Failure Mode simply means that maintenance is not the answer to manage the Failure Mode. No scheduled maintenance does not appear under the safety and environmental consequences column because a failure management strategy other than no scheduled maintenance must be identified for Failure Modes with safety or environmental consequences.

10.5 Other Default Strategies

A great many solutions other than proactive maintenance can be derived using the RCM process. The RCM Decision Diagram can lead a working group to identifying them. Examples of these include recommendations such as modifications to operating procedures, updates to technical publications, and equipment redesigns. In the context of RCM, these recommendations are known as Default Strategies.

The possibility of assigning a Default Strategy appears on the evident and hidden sides of the RCM Decision Diagram in numerous places, as depicted in Figure 10.10. Notice that even if a proactive task is identified as a solution, the working group still has the opportunity to assign a Default Strategy, which is a great strength of the RCM process. *RCM doesn't stop at formulating maintenance.* For example, there may be a component that is assigned a scheduled Replacement task. However, if the system were redesigned, the Replacement task could be eliminated. The working group can recommend the engineering redesign (a Default Strategy) *in addition* to recommending the Restoration task. However, it is important to note that working group members must identify failure management strategies that take care of assets in their current state. If a redesign or other change could improve the asset, the working group members still have to identify solutions that take care of the asset until the redesign is implemented.

Consider the example below:

Example Function: To deliver diesel fuel to the engine for continuous operation at an uninterrupted delivery pressure flow of 2,300-2,700 psi while operating under load.

Functional Failure: *Unable to deliver diesel fuel.*
Failure Mode: *Engine is serviced with incorrect fuel*
Excerpt from Failure Effect: *Within a very short time of starting the engine, the engine runs rough and begins to vibrate. The unit faults out for under frequency before the Operator can shut the unit down.*

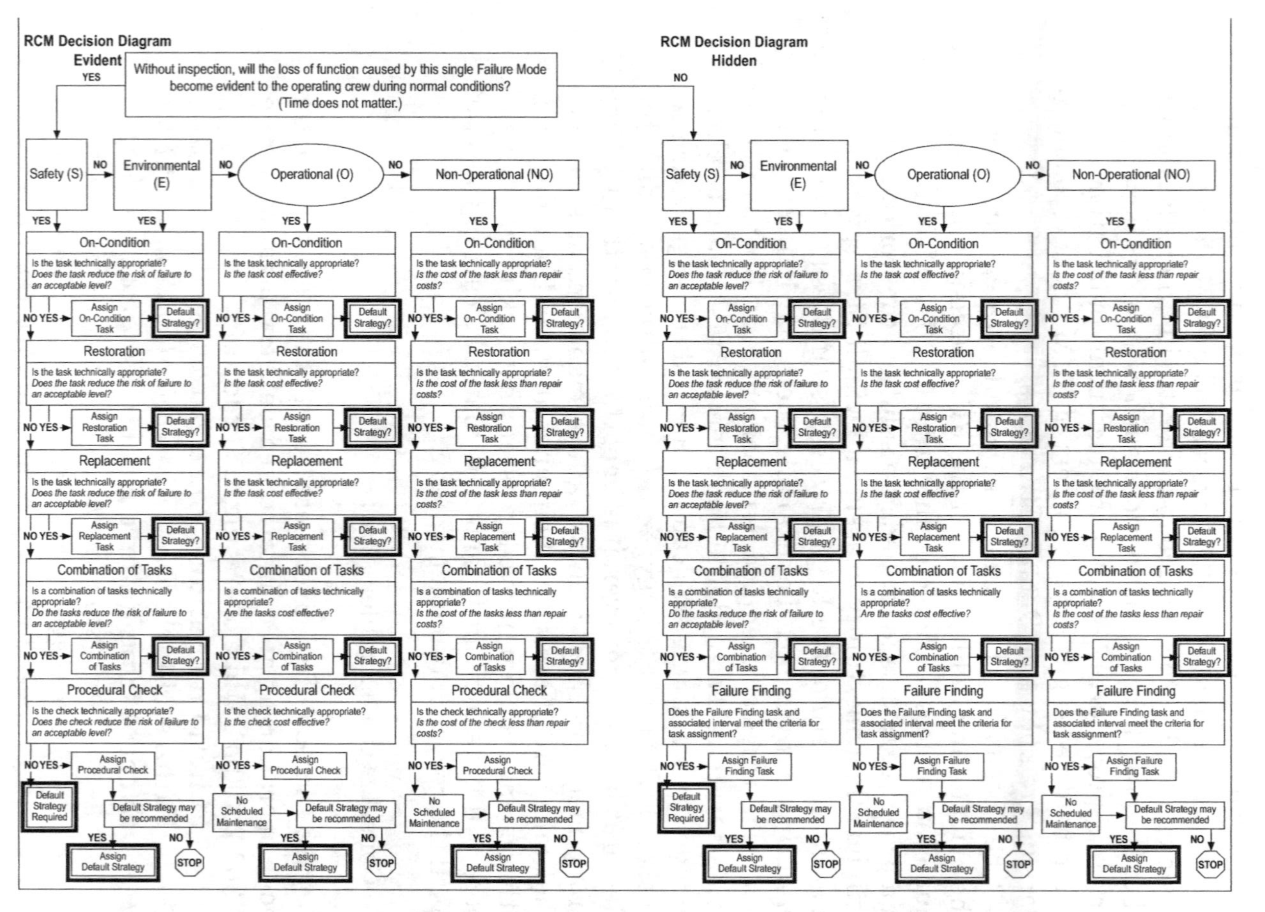

Fig. 10.10 Default Strategies represented on the RCM Decision Diagram

In this example, the working group has identified that the Failure Mode is evident and that it has operational consequences because the unit faults out. From there, the facilitator led the working group down the RCM Decision Diagram where the working group considered On-Condition, Restoration, and Replacement tasks, and a Procedural Check. None of these tasks were technically appropriate to manage the Failure Mode. The working group ended up in the *No scheduled maintenance* block. Notice on the RCM Decision Diagram depicted in Figure 10.10, after the *No scheduled maintenance* block, there is an opportunity to assign a default strategy. Although there is no task that is technically appropriate and worth doing, there is a solution that the working group recommends. It recommends the following Engineering Change: *Change the color of the fuel cap to prevent the engine from being serviced with the wrong type of fuel.*

Figure 10.11 illustrates how to document the above example on the RCM Decision Worksheet. The Failure Mode is evident so a *Y* is placed in the E column. The working group determined that servicing the engine with incorrect fuel would have operational consequences, so *N* is placed in the ES and EE blocks and a *Y* is placed in the EO column. The working group determined that there is no proactive maintenance task or Procedural Check that is technically appropriate and worth doing. Therefore, *N* is placed in the OC, Rst, RPL, C, and PC columns. However, the working group established that a Default Strategy would be appropriate to manage the Failure Mode. Therefore, *Y* is placed in the Default Strategy (DS) block and the recommendation is recorded in the Default Strategy column.

10.6 Important Notes Regarding Default Strategies

- As depicted in Figure 10.7, notice that Failure Finding tasks are considered last on the hidden side of the RCM Decision Diagram. Failure Finding tasks are considered last because any time a Failure Mode can be man-aged proactively, it should be. When a Failure Finding task is assigned, the organization accepts that the protective device may spend some time in a failed

RCM Decision Worksheet

Failure Mode			E	ES	EE	EO	ENO	HS	HE	HO	HNO	OC	Rst	Rpl	C	PC	FF	DS	Task	Initial Interval	Default Strategy
1	A	2	Y	N	N	Y						N	N	N	N	N		Y			Change the color of fuel cap to prevent the engine from being serviced with the wrong type of fuel.

Figure 10.11 Documenting a Default Strategy on the RCM Decision Worksheet

state. Therefore, it would be more advantageous to predict or prevent failure rather than to *find failures*. However, in most cases there isn't a proactive task that is technically appropriate for a protective device. Even in the case of the smoke detector, a scheduled replacement for the smoke detector battery can be scheduled every six months. However, this only manages the Failure Mode *battery discharges due to normal use*. The smoke detector's electrical circuitry cannot be *prevented* from failing. Therefore, when an organization assigns a Failure Finding task to blow smoke at the detector on a regular interval, it accepts the fact that the smoke detector may spend some time in a failed state between tests.

- When assigning Default Strategies such as engineering redesigns and technical publication updates, it is very important to remember that the working group is not there to redesign equipment or rewrite technical manuals. There simply is not enough time during an analysis to do so. This is yet another reason it is so important to have a trained facilitator lead the working group. The facilitator documents the Default Strategy but ensures that the working group does not spend excessive time detailing the solution. If working group members have specific ideas regarding a Default Strategy, the major points can be documented. Additionally, working group members can document their ideas off-line for inclusion in the validation package.

Summary

Assigning Default Strategies is the last step in the RCM process. Default Strategies include Procedural Checks, Failure Finding Tasks, no scheduled maintenance, and other Default Strategies such as engineering redesigns, operating procedure additions, and supply changes.

Procedural Checks are found only on the evident side of the RCM Decision Diagram. In the context of RCM, a Procedural Check is considered a Default Strategy because it isn't a proactive task. Instead, it is performed to check for a failure that may have already occurred because it would be unacceptable to operate, or continue to operate, with the failure condition. In the context of RCM, just like proactive tasks, Procedural Checks must be technically appropriate and worth doing in order to schedule the check.

Failure Finding tasks are the last type of task that is considered for hidden Failure Modes. A Failure Finding task checks if a protective device is in a failed state. A Failure Finding task is considered a Default Strategy because it isn't a proactive task; it seeks to *find a failure of a protective device* that has already occurred. The purpose of a Failure Finding task is to reduce the risk of a multiple failure to an acceptable level.

Assigning *no scheduled maintenance* to a Failure Mode simply means that maintenance is not the answer to manage the Failure Mode.

There are other Default Strategies that can be derived using the RCM process, such as modifications to operating procedures, updates to technical publications, and equipment redesigns.

Eleven

Analysis Validation and Implementation

Validation serves as Quality Assurance (QA) for each analysis.

Upon completion of an RCM analysis, it should be evaluated by a validation team that serves as Quality Assurance (QA) for each analysis. The purpose of the validation is threefold:

- Provides a review of the working group's decisions, to ensure the decisions are safe and technically defensible
- Brings chronic problems and their solutions to light by elevating them to management.
- Determines what recommendations will be implemented, prioritizes them, and sets the implementation process in motion.

11.1 Frequently-Asked Questions Regarding the Validation Process

When is a Validation Performed and How Long Does It Last?

Typically, a validation is scheduled to be performed within a few weeks of analysis completion. Validations take from two hours to several days to complete. However, the duration depends on the type, size, and complexity of the analysis. For example, a validation for an RCM analysis that contains 100 Failure Modes, on average, lasts 4-to-8 hours.

A validation is scheduled to be performed within a few weeks of analysis completion.

Who Makes Up the Validation Team?

The people who should attend the validation meeting are:

- Those— such as system engineers, maintenance managers, and manufacturing representatives—who have the technical knowledge to sanity check the decisions
- Management (or similar personnel) who have the authority to approve implementation of the results
- Other key members who will implement the results; these members may include representatives from tech nical publications, supply, and maintenance planning
- One-to-a-few working group members. No one can explain and support the decision process better than working group members themselves. Additionally, their participation allows consensus to be reached between the validation team and the working group with minimal delay.

Who Prepares the Validation Package and What is Included in It?

Upon completion of an analysis, the facilitator assembles the validation package.

Upon completion of an analysis, the facilitator assembles the validation package. The validation package consists of all the reports required for the analysis to be validated. Depending on the size and complexity of the analysis, it typically takes about a day to prepare and print validation packages (if hard copies are desired). The contents of a validation package are depicted in Figure 11.1.

Section	Description
Executive Summary	A short document (1-2 pages) that summarizes the RCM analysis.
Validation Flags	Details issues that require review during the validation meeting
Operating Context	Details the scope of analysis, the equipment, and the environment in which it is intended to operate
Information Worksheet	Includes Functions, Functional Failures, Failure Modes, and Failure Effects
Failure Mode Overview	A report specific to each Failure Mode and includes all data from the Information Worksheet in addition to data such as the decision logic, proposed maintenance, and default strategies
Proposed Maintenance Tasks	Details the proactive maintenance tasks recommended as a result of the RCM analysis
Proposed Default Strategies	Details the Default Strategies recommended as a result of the RCM analysis
Miscellaneous Documents	May include technical documents used during the analysis, notes section, cost analysis, etc.

Figure 11.1 Contents of an RCM Validation Package

How Is a Validation Conducted?

During the validation, an implementation strategy is formulated.

The analysis facilitator leads the validation meeting. During the validation, the details of each Failure Mode are reviewed and an implementation strategy is formulated.

There are various methods for conducting a validation. The following discussion outlines three approaches that are commonly used—the one that best accommodates the organization should be employed.

1. **The facilitator distributes the validation package to validation team members approximately one week before the validation.**

   This method shortens the duration of the validation meeting. Validation team members review the material ahead of time and have the opportunity to research various issues and formulate questions prior to arriving. The validation begins immediately.

2. **Distribute the validation packages at the validation meeting and allow the first few hours for the validation team to independently review the packages prior to starting the meeting.**

   This method allows validation team members—who are unlikely to review the package ahead of time because of demanding schedules—to review the document and make notes of any issues they would like to discuss once the validation begins. Once the package has been independently reviewed, the validation begins.

3. **Review the validation package as a team.**

   Using this method, the facilitator guides the validation team through the analysis one Failure Mode at a time as each one is validated.

Table 11.1 Key Topics for the validation team to focus on during an RCM Validation

	Key Topics for the Validation Team to Review
Operating Context	Ensure correct technical content regarding the equipment.
	Review the scope of the analysis to ensure that analysis content fits within the boundaries of the scope.
	Review analysis notes as these often detail assumptions that support various analysis decisions.
Functions	Ensure performance standards are what the organization "requires" versus what the equipment is "designed" to do.
	Ensure all protective devices are included in the analysis.
	Ensure protective device functions are written to account for the multiple failure. *Incorrect: To be capable of relieving pressure.* *Correct: To be capable of relieving steam drum pressure in the event that the pressure exceeds 500 psi.*
Failure Modes	Ensure that all reasonably likely Failure Modes that could occur in the operating context are included in the analysis.
	Ensure all current maintenance tasks subject to reverse engineering have appropriate Failure Modes identified.
Failure Effects	Ensure each Failure Effect is technically correct for the associated Failure Mode. Note: The following example assumes the fuel filter has a bypass feature: Failure Mode: Fuel filter clogs *Wrong Failure Effect: Fuel flow to the engine is restricted. Engine stops running...* *Correct Failure Effect: When the filter pressure differential reaches nine psi, the bypass valve opens allowing fuel to continue flowing to the engine. Unfiltered fuel is delivered to the engine...*
	Ensure each Failure Effect includes the worst case scenario (while still remaining plausible).
Failure Consequences	Ensure that each Failure Consequence is appropriate for the associated Failure Mode/Multiple Failure.
On-Condition Tasks	Ensure that the P-F interval identified is defined by *how quickly* the failure occurs once it is detectable, *not* based on how often the failure occurs.
	Ensure that the task interval is less than the P-F interval. Ensure that the net P-F interval is long enough to manage the consequences of failure. 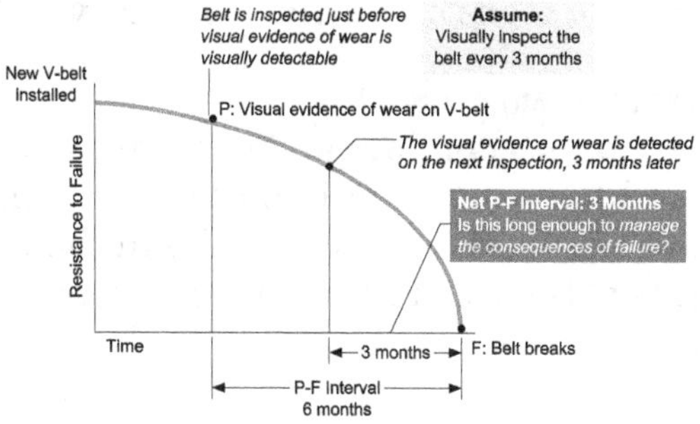 *Belt is inspected just before visual evidence of wear is visually detectable* **Assume:** Visually inspect the belt every 3 months New V-belt installed P: Visual evidence of wear on V-belt *The visual evidence of wear is detected on the next inspection, 3 months later* **Net P-F Interval: 3 Months** *Is this long enough to manage the consequences of failure?* Resistance to Failure Time 3 months F: Belt breaks P-F Interval 6 months
Replacement and Restoration tasks	Ensure that the *useful life* is used to assign task intervals—not the MTBF. 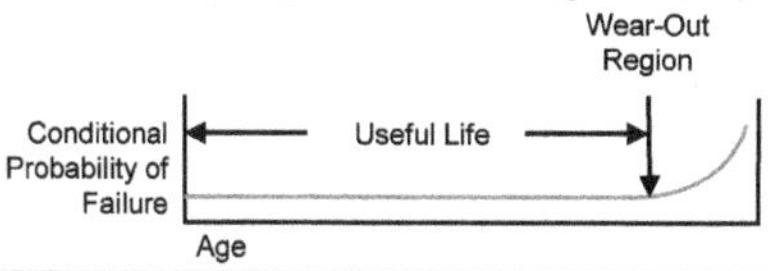 Wear-Out Region Conditional Probability of Failure Useful Life Age

continued on next page

Table 11.1 Key Topics for the validation team to focus on during an RCM Validation continued

Failure Finding Tasks	To the extent possible, ensure that each Failure Finding task checks the system as a whole. *Incorrect: Push the smoke alarm test button to ensure the alarm sounds.* *Correct: Using a spray can of smoke detector test aerosol, spray test material into the detector and wait 5 to 10 seconds for a response. Replace the smoke detector, as required.*
All tasks	Ensure that the synchronized task interval is equal to or less than the initial interval.
	Ensure that all tasks can be accomplished with the current equipment configuration and diagnostic tools available.
Synchronized Maintenance Task Intervals	Ensure the intervals for the final synchronized maintenance packages make sense and work for the organization.

On What Key Issues Should Validation Team Members Focus?

The validation team's review should ensure that each recommendation and its associated decision process are technically defensible. During the analysis, the facilitator creates *validation flags* for various issues that require the validation team's attention. Validation flags vary but can include issues on which the working group couldn't achieve consensus, areas where the working group requires additional information, or topics that required considerable discussion during the analysis. In addition to validation flags, Table 11.1 outlines key topics that the validation team should focus on during an RCM validation.

How are the Final Results Documented?

After the validation is complete and consensus between the validation team and the working group has been achieved, the facilitator prepares the final analysis results in the Final Summary. The contents of the RCM Final Summary are depicted in Figure 11.2.

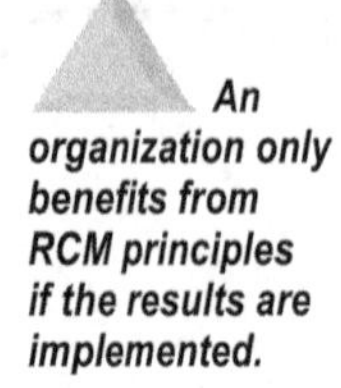

An organization only benefits from RCM principles if the results are implemented.

11.2 Implementing RCM Analysis Results

The validation team determines what recommendations will be implemented and prioritizes them, thus setting the implementation process

Section	Description
Executive Summary	A short document (1-2 pages) that summarizes the RCM analysis.
Operating Context	Details the scope of analysis, the equipment, and the environment in which it is intended to operate
Information Worksheet	Includes Functions, Functional Failures, Failure Modes, and Failure Effects
Failure Mode Overview	A report specific to each Failure Mode and includes all data from the Information Worksheet in addition to data such as the decision logic, proposed maintenance, and default strategies
Maintenance Tasks	Details the proactive maintenance tasks assigned as a result of the RCM analysis
Default Strategies	Details the Default Strategies assigned as a result of the RCM analysis

Figure 11.2 Contents of an RCM Final Summary

in motion. Of course, the results must be implemented before the organization can receive any benefit. An implementation plan should be developed and resources must be allocated to ensure its success. The details of such a plan vary from organization to organization and are beyond the scope and purpose of this book.

However, the obvious point is that an organization only benefits from RCM principles if the results are implemented. Even though most analyses are carried out well, many organizations that try RCM end up failing because the results are not implemented. Unfortunately, the RCM process is held responsible instead of the way the process was initiated. Implementation doesn't fail, for example, because maintenance planning doesn't have enough time to update the maintenance schedules or because maintenance personnel don't feel like changing their routines. The two main reasons why results aren't implemented, and why most RCM efforts fail, are:

- Buy-in from the organization from all applicable levels–*not just from the top*–was not obtained from the very beginning.
- The right people were not involved in the analysis and the validation process.

Having people caught unaware with changes they are responsible to implement, yet have no ownership in, is a recipe for failure. At times, validated recommendations for maintenance tasks and default strategies can be vastly different from what is currently in place. For example, simply presenting the maintenance planning department with seemingly drastic changes is inappropriate. They are likely to reject the changes because they don't understand them and because they have no ownership in them.

Chapter 12 outlines the steps required to ensure a successful RCM program. Included in those steps is fostering a dedication to the process. Even when RCM is carried out correctly, implementation can be difficult. *But when an organization is dedicated to RCM principles, implementation is just another step in the process—not a hurdle!*

Summary

Upon completion of an analysis, the facilitator assembles the validation package. The analysis is then evaluated by a validation team to ensure the working group's decisions are safe and technically defensible. The validation team also determines what recommendations will be implemented, prioritizes them, and sets the implementation process in motion. The final results of an analysis are detailed by the facilitator in the Final Summary and then passed on to the appropriate departments for implementation.

How to Initiate— and Successfully Sustain— an RCM Program

Starting an RCM program can seem overwhelming, especially if an organization doesn't have any first-hand experience with it. Many organizations that wish to implement RCM may have researched the process. They may understand that RCM can be used to transform maintenance and solve other problems, but don't know quite how to get started. The best way to implement a full-blown RCM program is to start off small. Take calculated steps that allow an organization to understand exactly how the process works, what it entails, how it is applied, the resources required to sustain it, and the benefits that RCM can provide. By building on the strength of initial results, an RCM program can be established and successfully sustained. The following steps outline how to accomplish exactly that.

12.1 The Steps to Initiate an RCM Program

1. Identify an RCM team leader.
2. Commit to applying RCM *correctly*.
3. Start off small.

4. Don't go it alone.
5. Obtain buy-in from the organization and spread the word throughout the organization.
6. Plan the pilot projects.
7. Conduct an introductory RCM training course.
8. Conduct the pilot project analysis.
9. Prepare the validation package.
10. Conduct the validation meeting.
11. Achieve consensus and deliver the Final Report.
12. Implement the results.

It is critical that someone within the organization leads the RCM effort.

Step 1: Identify an RCM Team Leader.

It is critical that someone within the organization leads the RCM effort. Very often, team leaders stand out by championing the effort from the beginning. RCM team leaders have many responsibilities. Two of the key responsibilities are:

- Lead the effort and ensure that momentum is maintained.
- "Sell" the process.

Lead the effort Team leaders have to take charge of the steps outlined here, especially during the early stages of RCM program initiation. In order to be effective, it is very important that the team leaders believe passionately about the process and can dedicate the time required to getting the RCM program off the ground and maintaining momentum.

"Sell" the Process Team leaders not only have to convince others of RCM's importance to the organization, but also must inspire individuals to believe in it. Team leaders should work with any outside RCM specialists who may be assisting with RCM implementation by echoing important points on behalf of the organization. Team leaders are often regarded as the "face of RCM" within the organization.

Step 2: Commit to Applying RCM *Correctly.*

Before an RCM program is initiated, it is important for an organization to commit to applying RCM correctly. This means that True RCM should be applied and the process should be carried out using a facilitated working group approach.

Although RCM is a resource intensive process, analyses can be completed efficiently if performed correctly with the right people.

Applying True RCM Many RCM processes are available. Therefore, choosing one is a very important step. It is often wrongly believed that RCM takes too much time and other resources to carry out. Although RCM is a resource intensive process, analyses can be completed efficiently if performed correctly with the right people.

True RCM analysis should be conducted by identifying the following, in the order in which they appear.

1. Functions
2. Functional Failures
3. Failure Modes
4. Failure Effects
5. Failure Consequences
6. Proactive Maintenance Tasks and Associated Intervals
7. Default Strategies

Streamlined RCM Caution should be taken when deciding which RCM process to use. Many streamlined RCM processes on the market boast fewer steps and shorter analysis time, claiming to achieve the same results. This simply is not true. Not only do these processes not produce the same results, they can be dangerous because they omit key steps in the process. Many streamlined versions of RCM differ significantly from what was intended by the original pioneers of the RCM process, Stanley Nowlan and Howard Heap. SAE JA1011, *Evaluation Criteria for Reliability-Centered Maintenance (RCM) Processes* describes these streamlined versions as "failing to achieve the goals of Nowlan and Heap and some are actively counterproductive."

To get the best results from RCM, analyses should be conducted with equipment experts—those who are most intimate with and knowledgeable of the asset and the operating environment in which the asset is expected to operate.

Using a facilitated working group approach to RCM

RCM applied correctly means the process is carried out *by the right people* using a facilitated working group approach. To get the best results from RCM, analyses should be conducted with equipment experts—*those who are most intimate with and knowledgeable of the asset and the operating environment in which the asset is expected to operate.* These equipment experts make up the RCM working group and represent various disciplines such as a mechanic, operator, original equipment manufacturer (OEM), and a systems engineer. The working group is led by a facilitator—an individual who is very well versed in RCM principles, leads the team, and ensures that RCM principles are being applied correctly. The team leader may elect to act as the facilitator.

Using this approach, the facilitator sits in front of the room facing the working group members. Figure 12.1 shows a working group assembled to analyze a boiler.

The facilitator asks the questions and working group members provide the technical information that is entered directly into the RCM database. The RCM database is digitally projected behind the facilitator so the working group members can view what the facilitator enters.

A facilitated working group approach is especially advantageous because the data required to answer all of the RCM questions often doesn't exist in current databases. Therefore, data collection is augmented by harnessing the knowledge and ex-

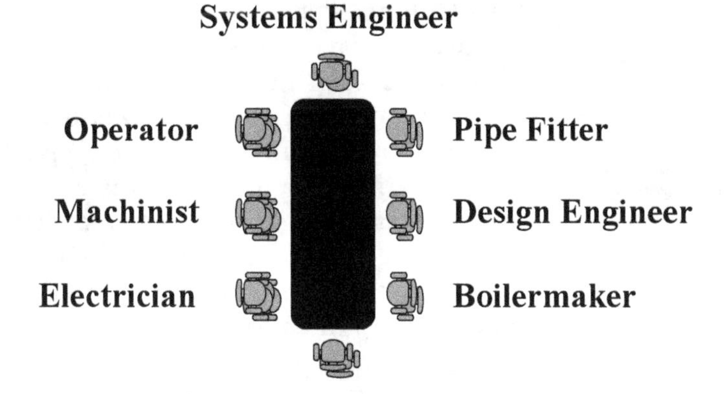

Figure 12.1 RCM working group assembled to analyze a boiler

Knowledge and expertise of working group members is the most truthful source of data.

pertise of working group members. In many cases, this is the most truthful source of data because a blend of experience, judgment, and operational data is utilized. This approach ensures that the system is analyzed from the perspective of what the organization *requires of the asset* (required performance) and what it *can* do (design capability). Working group members work together to dissect issues and formulate solutions. In many cases, issues are uncovered that are negatively contributing to the organization; these issues may otherwise have gone unchecked.

Step 3: Start Off Small.

The best and most logical way to initiate an RCM program is to start off small by commissioning a few pilot projects.

The best and most logical way to initiate an RCM program is to start off small by commissioning a few pilot projects—2 to 3 analyses are ideal. By choosing a few key systems to analyze, the organization can see exactly how the process works, what it entails, how it is applied, what resources are required to sustain it, and the benefits that RCM can provide. This allows an organization to make a minimum up-front investment. Pilot projects are discussed further in Step 6.

RCM is a particularly specialized process. Like any discipline, it takes time and experience to apply properly.

Step 4: Don't Go It Alone.

RCM is a particularly specialized process. Like any discipline, it takes time and experience to apply properly. If an organization doesn't have in-house RCM expertise when starting an RCM program, an outside RCM specialist should be enlisted to assist in successfully completing the pilot projects and conducting required training. RCM specialists can help in two ways.

- RCM specialists can provide required training to working group and validation team members and facilitate the pilot projects. This approach is the most efficient for completing the pilot projects and also requires the minimal investment.
- RCM specialists can provide required training to working group and validation team members. They can also provide RCM facilitator courses for in-house facilitators and then mentor the facilitators through the pilot projects. This method may not be ideal because an analysis typically takes twice as long to complete when facilitators are mentored than if experienced facilitators do it alone. If one of the goals of the pilot projects is to show how quickly RCM analyses can be completed, then this method is not recommended.

It is critically important to achieve buy-in from an organization before RCM is initiated—most especially from upper and middle management.

Step 5: Obtain Buy-In from the Organization and Spread the Word Throughout the Organization.

It is critically important to achieve buy-in from an organization *before* RCM is initiated—most especially from upper and middle management. It is vital to gain that approval before initiating RCM. Not achieving it is one of the biggest reasons organizations fail in implementing RCM.

Present a one-hour RCM brief to upper and middle management Upper and middle management can be briefed together or separately—whatever would be most appropriate for the organization. The RCM team leader should give the presentation. If an outside RCM specialist is involved, the RCM team leader and the outside RCM specialist can co-present or the RCM specialist can just be present to answer any questions that the team leader cannot.

Buy-in is critical from upper management because their support is often required to get the pilot projects off the ground. Their authority is essential when implementing RCM more fully within the organization. Perhaps most important, upper management typically makes the funding available to carry out RCM. Middle management is often left out of the process until their resources (RCM working group members!) are needed. However, leaving middle management out of the process in the beginning is one of the top reasons RCM initiatives fail. It is much easier to secure middle management's cooperation if they are involved from the outset.

Content of the one hour RCM brief The goals of the brief are for the attendees to understand the RCM process, recognize why it is in the best interest that it be applied within the organization, and to give permission (and funding!) to employ it. In order to achieve these goals, the brief should contain the following:

- *Briefly explain the history of RCM including a discussion on the six failure patterns.* (This discussion is especially helpful if the organization widely believes that most failures are more likely to occur as operating age increases.) Be clear that RCM is a time-honored, 40-year, proven process and is applied world-wide in the commercial and defense industries by private as well as government institutions. (It is not the project of the month!)

- *Define RCM and outline the seven steps.* Provide as much detail as is appropriate for the organization and the management level of the attendees. (For upper management, less is better!).

- *Show that RCM is conducted using a facilitated, working group approach.* Explain how much more powerful the process is when it is applied by equipment experts.

- *Describe the validation process.* Elaborate on the approval process.
- *Briefly outline how the pilot projects will be accomplished.* Explain that pilot projects will be finalized (discuss pilot project candidates, if appropriate), introductory training will be conducted, each pilot project will take one to two weeks, and then analyses will be validated. Be ready to share budget details.
- *Explain why RCM is advantageous to apply.* Detail what RCM has achieved for similar organizations. Stress that when it is applied correctly with the right people, it produces overwhelmingly positive results and can be used to enhance safety, reduce costs, improve availability, increase maintenance efficiency, improve environmental integrity, and achieve longer useful life of some components. If there are specific goals for applying RCM, explain them. (This is when it can be especially helpful to have the assistance of an RCM specialist. Most RCM specialists can provide specific success stories of other RCM analyses that have been completed within the organization's industry. Depending on the organization, it may be appropriate to have the RCM specialist at the briefing or to co-present the brief.) Justify to middle management why RCM requires them to commit their staff members for weeks to carry out RCM analyses.

Spread the word about RCM throughout the rest of the organization.

It is important to spread the word about RCM throughout the rest of the organization so others understand why RCM is being undertaken. Some organizations decide to conduct a large briefing. Others choose less formal channels such as publishing an article in the organization's newsletter or spreading the word through weekly team meetings. However the organization chooses to get the word out, the team leader should be involved.

The analysis planning process is incredibly important.

Step 6: Plan the Pilot Projects.

The analysis planning process is incredibly important. Well-planned analyses are far more successful than those that receive less attention. The following steps should be carried out by the facilitator and the team leader (the team leader may be the facilitator) while working with appropriate personnel (such as systems engineers, department leads, and members of management). However, analysis planning requires prior RCM experience. If an RCM specialist is training one of the organization's employees to facilitate during the pilot projects, the RCM specialist should assist in each step.

1. **Select the equipment subject to the pilot projects.**

   If there are many assets to choose from, select the equipment that is giving the organization the most pain. Advantageous systems to start with are those that may be experiencing chronic downtime, high operating costs, or high maintenance manhours. The key is to start with systems that offer the biggest chance of identifying areas of improvement. It's important that the pilot projects yield positive results so the strength of RCM can be demonstrated. A site visit by the facilitator is helpful, if it is possible, to examine the equipment first hand and experience the operating environment.

2. **Identify the goals and objectives of the pilot projects and delineate how success will be measured.**

   Goals and objectives vary from analysis to analysis. Examples of goals may be to establish a proactive maintenance plan, sanity check the current proactive maintenance plan, identify why equipment reliability is low, reduce operating costs, develop troubleshooting procedures, revise/augment/establish operating procedures, or identify training deficiencies. It should be determined ahead of time how success will be measured. Whatever means used, it is important that baseline data is established so results can be measured. For example, if the goal is to reduce operating costs, the current operating costs should be documented so that after the RCM analysis is completed, operating costs as a result of the RCM recommendations can be estimated. Upon implementation of the results, real-world operating costs can be tracked.

3. **Establish the scope of each analysis, determine how long each analysis will take, and identify how analysis sessions will be structured.**

   Boundary descriptions should specifically note what systems and components are included and excluded. Once this information is determined, the length of analysis sessions can be planned. (A skilled RCM specialist can very quickly estimate the time required to complete the analyses by walking down the equipment, reviewing technical manuals, and examining analysis goals.) Typically, analyses should be planned so that the analysis portion can be completed within a maximum of two weeks (cumulative time) using the facilitated working group approach. When identifying how long the analysis will take to complete, controversial or technically-detailed issues that may be involved with the equipment should be taken into consideration because they may add to required analysis time.

   For pilot projects, organizations typically choose to hold RCM analyses eight hours per day in order to complete the analyses as quickly as possible. However, analysis sessions should be scheduled to meet the organization's needs. (For example, analyses could be conducted three days per week for eight hours per day, or five days per week for four hours per day. However, conducting analyses this way extends the completion date of the projects. Usually it is desirable to complete pilot projects as quickly as possible to show how efficiently RCM can be done.)

4. **Identify working group and validation team members by name and discipline.**

   The experts associated with the asset that are considered vital for achieving the identified goals should be assembled to make up the working group. These experts may include representatives from operations, maintenance, engineering, design, technical publications,

Analysis results will only be as good as the input that is used to populate the RCM database.

logistics, safety, production, manufacturing, or any other specialty. Sometimes working group members are retained on an "on-call" basis. For example, a vibration or oil analysis expert may be needed to address only certain issues; therefore, full-time participation isn't necessary. It is imperative that people with the most experience—and thus people who have intimate knowledge of the system in its operating environment—are chosen to participate. Their participation is often a challenge because their expertise is needed in the field. It is important to note that analysis results will only be as good as the input that is used to populate the RCM database.

The validation team should be made up of people who know the equipment well enough to perform a technical review of the RCM analysis including the information populated in the database and the recommendations to ensure safety and technical defensibility. This team is usually comprised of individuals with technical experience as well as managers who are ultimately responsible for implementation of the results.

5. **Set the dates and venues of the introductory training courses, analyses, and validation meetings.**

- Setting the Dates

If possible, coordinate the dates of the training course so that the working group and validation team members for all pilot projects can attend. (More than one training course may be required depending on the number of delegates that require training.) The analysis and validation meeting dates should be chosen to accommodate the length of time required. The dates should be identified as quickly as possible so that working group and validation team members can be given adequate notice to commit to attending.

- Venue

Venues for working group sessions and validation meetings should be carefully chosen. The meeting room should be large enough to accommodate the working group members during the analysis and the validation team members during the validation meeting. It should also allow team members to sit in a U-shape formation as depicted in Figure 12.2.

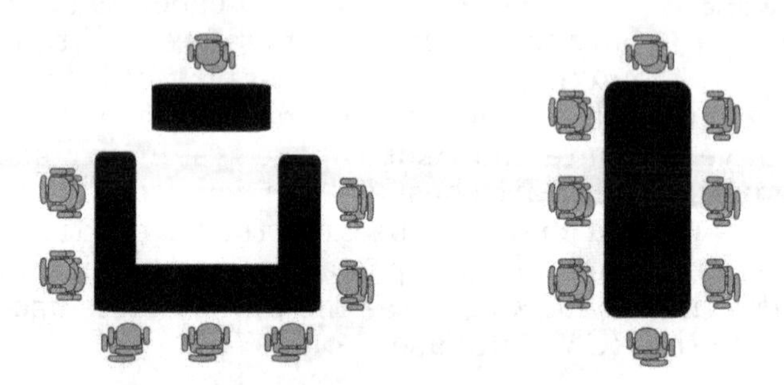

Figure 12.2 Meeting room layout for analysis and validation meetings

U-shape seating is essential as it permits all team members to make eye contact and enables the facilitator's ability to observe all team members; this seating in turn fosters a team environment. Classroom-style seating is not recommended as it does not allow for a cohesive working group approach.

6. **Identify if there are any obstacles that could hinder successful completion and implementation; formulate solutions.**

   To ensure successful completion and implementation of the analysis, any obstacles that could hinder progress should be identified and solutions should be formulated. These issues usually come to light during the facilitator's research/data gathering phase of planning.

7. **Notify working group members, validation team members, and supervisors in sufficient time to obtain commitment for participation.**

   Consistent attendance from working group and validation team members is absolutely critical to ensuring a technically defensible and successful analysis. When possible,

Consistent attendance from working group and validation team members is absolutely critical to ensuring a technically defensible and successful analysis.

team members should be notified at least four weeks in advance that their attendance is required to ensure that people have enough time to plan to attend. It is also important to notify team members' respective supervisors to ensure commitment for participation. If a response is not received within a few days, follow-up should be performed. Additionally, a reminder should be issued one week before the analysis/validation meeting. Scheduling RCM training, analyses, and validation meetings requires considerable work, so someone who has the time to do so should be assigned this responsibility. (During the pilot project phase, the team leader often performs this duty. However, if time does not permit, this responsibility may be delegated.)

8. **Gather technical documentation and facilitation materials required for the analysis.**

Various technical documents are referenced during the analysis and validation, so any materials that are required should be obtained prior to starting the analysis. Items that should be considered are technical manuals, current maintenance plans, operating procedures, emergency procedures, drawings/schematics, washing and cleaning procedures, memorandums, photos, pre-op and post-op inspections, startup and shutdown procedures, historical data, etc.

During an analysis and validation meeting, the facilitator is positioned in front of the room and inputs the data received from the working group/validation team into the RCM database, which is projected on a screen behind the facilitator. Additionally, flip charts or a white board are occasionally used to sketch ideas and schematics during the analysis/validation. As such, the facilitator must ensure that a laptop computer, digital projector, flip charts/white board, markers, and a laser pointer are available.

Step 7: Conduct an Introductory Training Course.

Understanding the basic principles of RCM analysis is vital to the working group and validation team members' successful participation.

All working group and validation team members must attend an RCM introductory training course. This course should be presented by an RCM specialist who can teach the material and field questions appropriately. Understanding the basic principles of RCM analysis is vital to the working group and the validation team members' successful participation. It allows them to understand the RCM specific questions the facilitator poses. (For example: *What is the P-F Interval? What is the useful life?*) Validation team members need to be well versed in RCM principles to review and validate the results. Keep a running list of the people who have been trained and include the training date, discipline (e.g., outside machinist, operator), and the analyses in which they've participated.

Step 8: Conduct the Pilot Project Analysis.

The facilitator leads the working group by asking questions during the analysis. The working group members provide the technical information. All of the information gathered during the RCM analysis is entered directly into an RCM database as the analysis is being conducted. The facilitator works to achieve consensus amongst team members on all issues.

Step 9: Prepare the Validation Package.

Upon completion of the analysis, the facilitator prepares the validation package. The validation package should contain all the documentation necessary to perform a technical review of the working group's recommendations. The contents of the validation package are depicted in Figure 11.1.

Step 10: Conduct the Validation Meeting.

Upon completion, each analysis is evaluated to ensure technical defensibility by the validation team.

Upon completion, each analysis is evaluated to ensure technical defensibility by the validation team. The validation meeting is conducted by the facilitator. Any technical content that the validation team questions or any additions they wish to make to the analysis are detailed in the RCM database so that consensus can be achieved with the working group. Analyses are an effective way of identifying problems and proposing solutions. They also elevate issues to the appropriate level of management so that changes can be implemented. The tasks required to implement analysis results are recorded in the form of action items. Each action item is assigned a due date and delegated to an appropriate person.

Step 11: Achieve Consensus and Deliver the Final Report.

Achieving consensus regarding all matters is one of the most important responsibilities of an RCM facilitator. Consensus means that all team members involved agree with the information and decisions recorded in the RCM database. Therefore, if there is information that the validation team wants to add to the analysis or recorded decisions that they wish to challenge, the RCM facilitator must achieve consensus with the working group members. The RCM facilitator prepares the Final Report after consensus is reached.

Step 12: Implement the Results.

After the RCM analysis has been validated, the results must be implemented before the organization can receive any benefit.

After the RCM analysis has been validated, the results must be implemented before the organization can receive any benefit. The results can include the elimination of current maintenance tasks, the implementation of new ones, current task intervals being shortened or extended, redesigning hardware, and/or modifying operating procedures. An implementation plan should be developed and resources must be allocated to ensure its success. The details of such a plan vary from organization to organization.

12.2 Sustaining an RCM Program

Once the pilot projects have been completed, the organization can assess the results and decide if RCM is worth fully implementing.

Once the pilot projects have been completed, the organization can assess the results and decide if RCM is worth fully implementing. If so, adequate time and resources must be directed towards the effort. Successfully establishing and sustaining an RCM program are dependent on fulfilling the following.

1. Establishing dedicated RCM team members.
2. Providing proper training for all RCM team members.
3. Implementing a scheduling process and adhering to it.
4. Establishing a reporting system to communicate results and program status.

Establishing Dedicated RCM Team Members

Successfully sustaining an RCM program takes considerable coordination and cooperation and, thus, requires dedicated team members.

Successfully sustaining an RCM program takes considerable coordination and cooperation and, thus, requires dedicated team members. Oftentimes when an organization elects to implement RCM, the resources required to successfully sustain it are underestimated. Establishing the required resources is a key component of maintaining an RCM program.

There are several roles and associated responsibilities that must be filled in order for the program to function efficiently and effectively. The key roles are the project director, RCM team leader, RCM coordinator, and the facilitator. Every position plays an integral role in the overall process and all team members must work together in order achieve success. It is critical that RCM team members not only believe in the process, but also have enough time to dedicate to their respective duties to support it.

The number of RCM team members required depends on the number of analyses an organization plans to perform in a year. The following discussion assumes that eight analyses, each requiring two weeks, will be accomplished. For the first year of a program, this goal is a reasonable one for an organization that has successfully completed a few pilot projects.

Project Director The project director oversees the RCM program. Any problems that arise which cannot be solved by the RCM team leader should be elevated to the project director. In general, the project director obtains and manages funding for the program and oversees the RCM team leader. Typically the Project Director isn't required full time. On average, four hours per week are spent on RCM responsibilities.

The RCM team leader runs the RCM program and is the "face" of RCM.

RCM Team Leader The RCM team leader runs the RCM program and is the "face" of RCM. Therefore, it is important that the RCM team leader has RCM experience. It is highly recommended that the RCM team leader served previously as an RCM facilitator. This experience provides instant credibility and allows smooth transition into a full-blown program.

The RCM team leader works with management to select the systems to be analyzed, identifies the facilitator (when more than one facilitator is assigned to the RCM program), meets with the RCM coordinator to delegate and supervise planning, and is the first in the chain of command when difficulties arise. The team leader also oversees distribution of the Final Package to the appropriate personnel/department for implementation. Another responsibility is to assemble presentations summarizing analysis results and to brief the results, as required. If the RCM team leader also acts as the facilitator, this is a full-time job and requires the assistance of a full-time RCM coordinator. The RCM team leader reports directly to the RCM project director.

RCM Coordinator The RCM coordinator is responsible for organizing and scheduling analyses. Significant planning and follow-up work are required for analyses to be successful. Although strong organizational and administrative skills are necessary, the role of an RCM coordinator does not require a technical background.

The RCM coordinator works with the facilitator and also interfaces with the RCM team leader. Specifically, the RCM coordinator is responsible for the following:

- Working with the RCM team leader and facilitator, maintains the schedule for training courses, analyses, and validation meetings; documents which delegates require the introductory training course, and records the associated contact information

- Organizes a meeting with the facilitator and system engineer (or appropriate personnel) for technical planning once a system has been identified for RCM analysis
- Identifies and reserves the conference room for the analysis and validation meeting
- Contacts the facilitator to report analysis and validation meeting venues
- Assists the facilitator in gathering/reserving presentation materials required for analysis
- Notifies and follows up with working group and validation team members (and their supervisors when appropriate) to obtain commitment for participation
- Assists the facilitator in preparing and distributing validation packages and Final Packages
- Supports the facilitator with other duties, as required

Although the RCM Coordinator works directly and extensively with the facilitator, the RCM Coordinator reports directly to the RCM team leader. The RCM coordinator is a full-time position.

An RCM facilitator is a specialist who has a thorough understanding of, and a sincere belief in, the basic principles of RCM.

RCM Facilitator Facilitation, according to the American Heritage Dictionary, is *the act of making a process easier*. In the context of RCM, facilitation is a practice by which a facilitator leads a team of equipment experts, the working group, and validation team—thus allowing each team to focus on the formulation of technically appropriate, safe, and cost-effective decisions regarding the asset being analyzed. An RCM facilitator is a specialist who has a thorough understanding of, and a sincere belief in, the basic principles of RCM. Learning to facilitate is not a simple matter or something that can be learned by merely reading a book. Facilitation takes time and practice to master and requires many skills and abilities.

The RCM facilitator is the most important person in the RCM process and should be chosen carefully. The following discussion outlines the key responsibilities, work and education requirements, and skills and abilities required of a facilitator.

Key responsibilities of a facilitator

1. Plans the RCM analysis
2. Facilitates the RCM analysis
3. Produces the RCM validation package
4. Facilitates the RCM validation meeting
5. Achieves consensus amongst working group and validation team members
6. Produces the Final Report
7. Assembles a presentation summarizing analysis results and briefs the results, as required

Work and Education Experience Requirements

RCM facilitators must have a strong technical background such as an engineering degree or commensurate technical experience. This is required for three main reasons: 1) allows facilitators to understand the RCM process and how it applies to physical assets; 2) lays the foundation for professional credibility with the working group and validation team; and 3) allows facilitators to discern technical issues and competently ask the technically appropriate questions required to identify how each issue should be handled in the context of RCM.

Skills and Abilities

Facilitating an RCM analysis requires a wide range of skills and abilities. Effective facilitators must embody the following:

- **Believe:** When applied correctly, the basic principles of RCM make a unique and vital contribution to physical asset management. Facilitators must be energized by and wholeheartedly embrace these principles.

- **Teamwork:** Energized by teamwork and can cooperate with others to meet objectives.

- **Leading Others:** Set goals and functions like a field marshal with the ability to strategically organize, motivate, and inspire team members to accomplish these goals in a timely manner. At the same time, create and maintain a sense of high morale, order, and direction. Ensure that the team members are applying RCM correctly and making essential corrections as required.

- **Interpersonal Skills:** Conduct themselves and inter act with team members such that they are respected, liked, and considered technically credible. Facilitators must be able to motivate and lead their teams, while developing a sense of camaraderie.

- **Diplomacy and Tact:** Facilitators will encounter team members with many different personalities and dispositions in addition to varying levels of education, competence, and responsibility. These differences must be managed appropriately. Additionally, facilitators must be able to effectively manage discord amongst team members in a kind and professional manner, while maintaining control.

- **Objective Listening:** Must be sufficiently objective to receive and understand numerous and varying inputs from team members.

- **Empathetic Outlook:** Ability to understand the various perspectives of each team member and recognize how each of these may affect the organization differently. While facilitating, must identify with what they are hearing and observing and understand that each unique perspective is important to that team member in a different way.

- **Developing and Influencing Others:** The capacity to contribute to the development and growth of team members and to personally influence their opinions, thinking, actions, and decisions.
- **Conflict Management:** The capacity to effectively resolve differing points of view and gain consensus amongst team members.
- **Planning and Organization:** The ability to create effective strategies and implement tactics to achieve strategic goals.
- **Writing Skills:** Must be able to summarize comments from team members in a succinct form for inclusion in the analysis.
- **Computer Skills:** Must have efficient computer and keyboarding skills.

A Facilitator's Workload

Facilitation is a full-time position. Each analysis cumulatively takes approximately five weeks of a facilitator's time to complete. (Of course, this time may vary slightly from analysis to analysis.) The five weeks are distributed as follows: seven days planning the analysis and preparing to facilitate; ten days facilitating the analysis; three days reviewing, preparing, and distributing the validation package; one day conducting the validation meeting; two days achieving consensus on the results; and two days preparing and distributing the Final Package. The RCM facilitator reports directly to the RCM team leader (if there is a facilitator under the RCM team leader).

Everyone involved in the RCM process should be properly trained according to their level of participation.

Training for All RCM Team Members.

Participants then understand the process and how they can contribute to the success of RCM. The following discussion outlines the various RCM training opportunities.

Introductory presentation The introductory presentation typically lasts one hour. The presentation includes a very high-level discussion of the RCM process, an explanation of the major tenets, and details how the RCM process will benefit the organization.

Introductory course The introductory course typically lasts two days. It provides basic knowledge of the RCM process. Specifically, the introductory course covers a discussion on:

- Introduction to RCM
- Failure data and maintenance
- The seven steps of the RCM process
- Task synchronization
- RCM validation process

RCM Facilitator Course The RCM facilitator course typically lasts two weeks. This classroom instruction provides an in-depth understanding of RCM principles and how to apply them. Additionally, the course helps to develop the skills needed in order to effectively lead a team and to gain consensus. Specifically, the facilitator course includes instruction on how to:

- Plan the RCM analysis
- Facilitate the RCM analysis
- Produce the RCM validation package
- Facilitate the RCM validation meeting
- Produce the final report
- Assemble and give a presentation summarizing analysis results

RCM Facilitator Mentoring In addition to classroom training, all facilitators must complete a period of mentoring to transform their theoretical knowledge to the practical skills required to produce safe and technically defensible RCM analyses. How

Table 12.1 Summary of the Training Required for RCM Team Members

	Project Director	Team Leader	Facilitator	RCM Coordinator	Management	Working Group Members
Introductory presentation	✓	✓	✓	✓	✓	as required
Introductory course	✓	✓	✓	✓	as required	✓
Facilitator course	*	*	✓	N/A	N/A	N/A
Facilitator mentoring	N/A	N/A	✓	N/A	N/A	N/A

* *Some organizations find it advantageous for project directors and team leaders (when the team leader isn't acting as the facilitator) to have more RCM knowledge and expertise than the introductory course provides. In these cases, the project director and the team leader can attend a facilitator course.*

much mentoring each facilitator requires varies from individual to individual.

Table 12.1 summarizes the level of training required for RCM team members.

Some organizations find it advantageous for project directors and team leaders (when the team leader isn't acting as the facilitator) to have more RCM knowledge and expertise than the introductory course provides. In these cases, the project director and the team leader can attend a facilitator course.

Implementing a Scheduling Process and Adhering to It

This step is one of the most important responsibilities of RCM program management because ensuring delegates arrive at their respective meetings is critical. A scheduling system appropriate for the organization must be established and maintained by the RCM coordinator. Accountability should be firmly upheld.

One of the biggest mistakes an RCM coordinator can make is simply sending out an e-mail that invites a working group or validation team member to an analysis or validation meeting. Sending an e-mail can be considered the *first* step in the notification process, but it must be *followed* with a phone call or visit to confirm attendance. Delegates' supervisors should also be notified (if this is an appropriate step for the organization). Even after a delegate has responded positively, one week before the analysis the RCM Coordinator should contact all delegates again to confirm participation. Dates of initial notification, personal follow up, and one week confirmation should be tracked.

Establishing a Reporting System to Communicate Results and Program Status

RCM results must continually be reported to appropriate personnel so that RCM success is communicated throughout the organization. How much information is communicated and what means is used to disseminate the information should be appropriate for the organization.

Summary

Starting an RCM program can seem overwhelming, especially if an organization doesn't have any first-hand experience with it. The best way to implement a full-blown RCM program is to start off small and take calculated steps that allow an organization to understand exactly how the process works, what it entails, how it is applied, what resources are required to sustain it, and the benefits that RCM can provide. By following the steps outlined in this chapter, establishing dedicated RCM team members, and providing proper training to them, a successful RCM program can be initiated and sustained.

Frequently Asked Questions and Common Misconceptions Regarding RCM

The following discussion outlines Frequently Asked Questions (FAQs) regarding the RCM process in addition to common misconceptions of the process.

13.1 RCM FAQs

At What Point in an Asset's Lifecycle Should RCM be Applied?

RCM should be applied as early in an asset's lifecycle as possible—beginning with the design phase.

RCM should be applied as early in an asset's lifecycle as possible—beginning with the design phase. As detailed in Chapter 1, Section 1.8, it is critically important to define what is required of an asset. In the context of RCM, writing Functions allows an organization to document *specifically* what is required of an asset so that the RCM team can determine if the asset is capable of serving in that manner. There are countless examples of equipment that make it to the field but are unable to provide the required performance which, worst case, can be deadly and, best case, requires redesign after the asset if fielded. Additionally, applying RCM during the design phase allows Failure Modes to be identified that may have resulted in expensive or difficult proactive maintenance requirements.

RCM can be applied at any point in the system's lif e cycle, resulting in tremendous benefits.

However, there are many active and successful legacy RCM programs across the world. Many were initiated decades after the system was fielded. Because the RCM process is zero-based, applying RCM to legacy equipment allows years of tradition and bad habits to be sanity checked so that the most safe and cost-effective failure management strategies can be implemented. RCM can be applied at any point in the system's lifecycle, resulting in tremendous benefits.

What Role Does Software Play in the RCM Process?

When RCM principles were first developed, the process was largely paper-driven. However, today's technology allows robust RCM software applications to be designed. There is so much information generated during an RCM analysis that it would be absurd to try and manage all of it without a software product specifically designed for RCM analysis. Personal experience dictates that trying to keep track of the information gathered during an RCM analysis using a basic spreadsheet program is incredibly confusing and not worth the effort. The smartest way to organize RCM data is to use a software program specifically designed to do so. However, remember that any RCM software product should be used to store the information generated by RCM analysis—*the RCM working group makes the decisions.*

Once RCM Analyses are Completed, Are They Ever Revisited?

Yes, RCM is considered a living program. RCM analyses should be reviewed periodically but there is no fixed interval at which analyses should be reviewed. The frequency should be dictated by how often equipment circumstances change such as operational tempo, operating environment, and changes in required performance. During a review of an RCM analysis, issues such as revising existing Functions, identifying new Failure Modes, modifying proactive task frequencies, and identifying areas for technology insertion can be accomplished. Depending upon the number of circumstances that have changed since the original analysis was completed, reviewing an RCM analysis may take just a few hours or could take up to several days. Just like the original analysis, review of the RCM analysis should be accomplished using a facilitated working group approach.

Can RCM Templates Be Used in Order to Expedite an RCM Analysis?

Great care must be taken if any kind of a template is used during an RCM analysis such as FMECA templates, Failure Mode libraries, and complete RCM analysis templates. Oftentimes, the template does not include an explanation of the equipment operating context on which the predetermined data is based. For example, an RCM analysis of a compressor system used in two different operating contexts can vary significantly.

Additionally, Failure Modes included in Failure Mode libraries are often written at a very low level. Considerable time is required to determine if any are applicable or to rewrite relevant Failure Modes at a higher level so that an appropriate failure management strategy can be formulated. (Writing Failure Modes at varying levels is discussed in Chapter 6, Section 6.5.) Using templated information can result in an analysis taking much longer than if a working group generated the Failure Modes on its own from scratch. Therefore, using any kind of predetermined information should be approached with great caution, and then only if an experienced facilitator leads the working group.

Can the Information Included in an RCM Analysis Be Used for Other Purposes?

There is a vast amount of information documented during an RCM analysis. With minimal effort, the FMEA (Steps one through four of the RCM process) can be used to generate a troubleshooting guide or to augment current troubleshooting procedures.

13.2 Common Misconceptions Regarding the RCM Process

Misconception 1: RCM is Just About Deriving Proactive Maintenance.

One of the major products of an RCM analysis is the development of a proactive maintenance program. However, as discussed in Chapter 1, Section 1.4, and in Chapter 10, Section 10.5, RCM can be used to formulate other solutions such as new operating procedures, updates to technical publications, modifications to training programs, equipment redesigns, supply changes, enhanced troubleshooting procedures, and revised emergency procedures. When a facilitated working group approach to RCM is used, equipment experts can identify a system's vulnerabilities and formulate failure management strategies to manage them.

Misconception 2: RCM is a Process that is Used to Reduce Maintenance.

RCM is often regarded as a process that is used to reduce maintenance. Maintenance is often reduced when RCM is applied to a legacy asset whose scheduled maintenance plan was formulated with the incorrect assumption that the likelihood of most Failure Modes increases with increased operating age. However, RCM results from analysis to analysis vary greatly. Some RCM analyses may increase the amount of scheduled maintenance that is performed but may result in improved equipment reliability. It is important to remember that the intention of RCM is to formulate failure management strategies that are technically appropriate and worth doing—not to reduce maintenance.

Misconception 3: RCM has Serious Practical Weaknesses in an Industrial Environment.

If any organization finds weakness in the RCM process, then the weakness must reside in its application—not the process. The principles of RCM are so robust and so universal that they can be applied—*and are being applied*—in nearly every industry throughout the world. Chapter 12 outlines how to initiate and successfully sustain an RCM Program.

Misconception 4: Reading a book or attending an introductory course enables an individual to implement RCM on his own.

Misconception 5: RCM is Too Time and Resource Intensive.

Misconceptions 4 and 5 share the same explanation. RCM is a particularly specialized process. Like any discipline, it takes time and experience to apply properly. If an organization doesn't have in-house RCM expertise when starting an RCM program, an outside RCM specialist should be enlisted to assist in successfully initiating the program. Not having properly-trained personnel carry out RCM leads to excessive analysis time and poor results. This is a major reason why the basic principles of RCM have been criticized and manipulated. The fault doesn't reside in the RCM process; the fault resides in its flawed application.

Misconception 6: RCM Seeks to Analyze Every Failure Mode

RCM does not seek to analyze every Failure Mode. Instead, RCM provides specific criteria for which Failure Modes should be included in an RCM analysis. Chapter 6, Sections 6.4 and 6.5 outline the four criteria for determining which Failure Modes should be included in an analysis as well as how detailed Failure Modes should be written, respectively.

Misconception 7: RCM is a Process that was Designed for Use on Aircraft—*Only.*

As explained in Chapter 1, Section 1.6, although RCM principles were first developed within the aviation industry and were included in MSG-1, MSG-2, and then finally MSG-3, the first time the term *"Reliability Centered Maintenance"* ever appeared was in Nowlan and Heap's groundbreaking book *Reliability-Centered Maintenance.* Furthermore, Nowlan and Heap intended for RCM to be used outside the aviation industry. In their book, they state, "This volume provides the first full discussion of Reliability-Centered Maintenance as a logical discipline for the development of scheduled maintenance programs."

Regarding RCM's rigor, Nowlan and Heap write, "The content of scheduled maintenance programs developed by experienced practitioners of MSG-2 techniques may be quite similar to the programs resulting from RCM analysis, but the RCM approach is more rigorous, and there should be much more confidence in its outcome.... The RCM technique can also be learned more quickly and is more readily applicable to complex equipment *other than transport aircraft."* RCM has been used and continues to be used throughout the world with incredible results. The principles of RCM are versatile and robust and can be used in nearly any industry.

Misconception 8: RCM, FMEA, and FMECA are Independent Processes.

Steps 1-4 of the RCM process (Functions, Functional Failures, Failure Modes, and Failure Effects) make up a Failure Modes and Effects Analysis (FMEA). Steps 1-5 of the RCM process (Functions, Functional Failures, Failure Modes, Failure Effects, and Failure Consequences) make up what is known as a FMECA. When RCM is performed, the requirement for a FMEA and FMECA is largely satisfied.

Misconception 9: Condition Based Maintenance (CBM) and RCM are Independent Processes.

Step 6 of the RCM process, *Proactive Maintenance and Associated Intervals,* embodies the process of considering Condition-Based Maintenance (On-Condition tasks). Therefore, if RCM is properly carried out, CBM is accomplished as a natural order of course.

Misconception 10: Working Group Members Guess When Answering Questions Posed During an RCM Analysis.

Because they are so intimate with the equipment and the operating environment, working group members are in an excellent position to answer the questions posed during an RCM analysis. They often use experience and judgment to provide answers, *but they don't take guesses.* A facilitator is trained to recognize when a working group doesn't know—and appropriate action is taken as a result. When the working group doesn't have the answer, the issue requiring additional information is parked until the answer can be found.

Fourteen

RCM is Only Part of the Solution

It has long been debated (and even argued) that True RCM is the best and most responsible type of analysis to apply to all assets. Indeed, RCM is a remarkable tool. If it is carried out correctly with the right people, amazing results can be produced. However, there are many who believe that True RCM is unnecessary because it takes too much time and other resources. It is often recommended that a shortened version of RCM (a streamlined RCM process as discussed in Chapter 1, Section 1.6) is just as effective, if not better. This just isn't so.

The use of streamlined RCM as a substitute for* True RCM *is strongly discouraged.

RCM alone is not a total solution. At times, there is a legitimate need to apply less robust processes to some assets.

The use of streamlined RCM as a substitute for True RCM is strongly discouraged. An organization that wishes to apply RCM should apply it correctly. However, personal experience and application have proven that RCM alone is not a total solution. At times, there is a legitimate need to apply less robust processes to some assets; these processes can nevertheless be entirely appropriate under certain circumstances. This chapter explains why RCM alone is not a total solution. It introduces *The RCM Solution*, which embodies additional analyses for managing assets.

Chapter 14

14.1 Two Fundamental Realities of Asset Management

The following discussion is intended to serve asset custodians in their pursuit of dealing responsibly with two fundamental realities, regardless of the complexity of what is being cared for. These realities are true for all assets such as aircraft, ships, ground vehicles, missiles, HVAC systems, and machine tools. The following discussion considers these realities in a manufacturing plant.

Reality 1:

There are thousands of assets across a plant, many of which need attention in the way of formulating failure management strategies.

These assets include:

- Major manufacturing systems such as dehumidifiers, compressor systems, chillers, and boilers
- Support equipment including tow tractors, mobile electric power plants, and mobile air conditioning units
- Machine tools such as band saws, pipe benders, table saws, hydraulic presses, and lathes
- Vehicle garages equipped with wheel balance machines and hydraulic lifts
- General support buildings which include carpentry and welding shops

Each of these assets was obtained for a particular reason and each serves the plant in different ways. Furthermore, they have varying:

- Responsibility with respect to mission criticalness
- Roles in plant production
- Levels of historical unreliability
- Degrees of technical difficulty

Reality 2:

Funding and other resources are pretty much never available to support a full-blown RCM analysis for *every* asset.

As a result, regardless of the complexity or nature of what is being cared for or how it serves the plant, the RCM process is often shortcut and a streamlined version of RCM is adopted as the analysis of choice in lieu of RCM. This poses a problem—and can even be dangerous—because some assets require the rigor of True RCM.

14.2 The Application of Processes Less Robust than RCM

In order to deal with the two realities discussed above, shortcut versions of RCM are often employed, but for the wrong reasons.

In order to deal with the two realities discussed above, shortcut versions of RCM are often employed, but for the wrong reasons. Although these realities may seem to offer justification to employ a less robust analysis than RCM, they are the wrong reasons and are not technically justifiable.

Why *Shouldn't* Responsible Custodians Apply Less Robust Processes?

Less robust processes should not be applied because they are quicker—*although they may be quicker*. Less robust processes should not be applied because there isn't enough funding for True RCM—*although there often isn't enough funding*. Less robust processes should not be applied because they offer the same results—*because they don't.*

When *Should* Responsible Custodians Apply Less Robust Processes?

Less robust processes should be applied when it has been determined that it is the right thing to do for a particular asset and the organization it serves.

14.3 The RCM Solution

So how does an equipment custodian determine when it is right to apply less robust processes and what level of rigor is appropriate for each asset or system? The following system, *The RCM Solution,* helps equipment custodians manage numerous assets and systems while managing the two fundamental realities described previously. The RCM Solution is depicted in Figure 14.1.

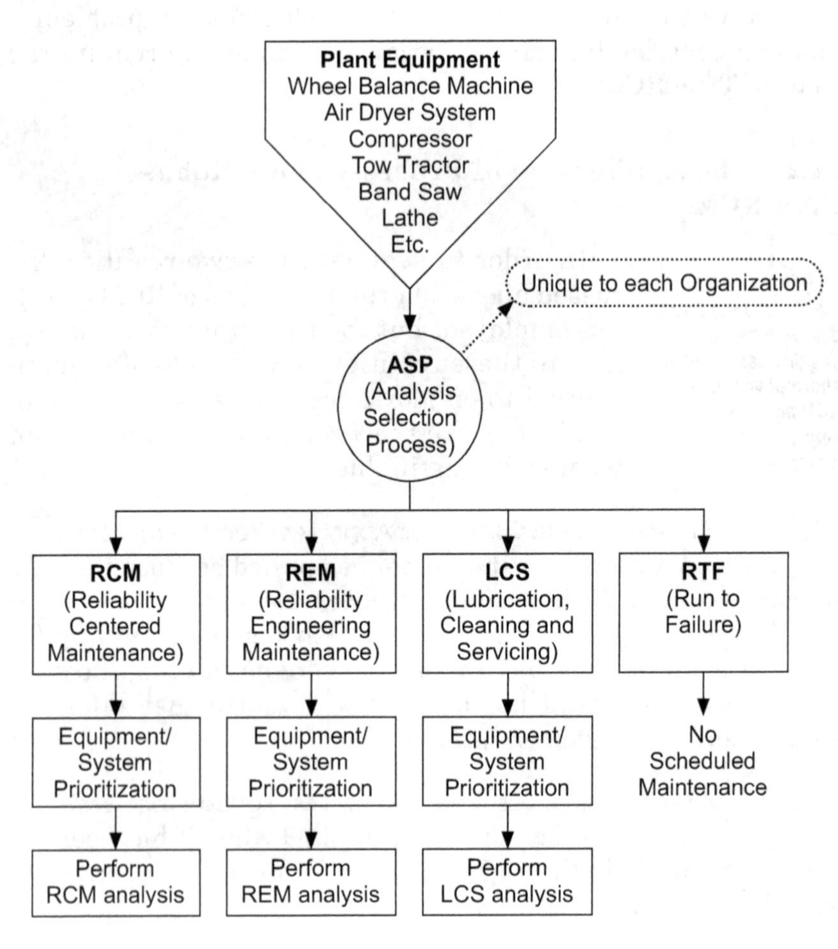

Figure 14.1 The RCM Solution

RCM is Only Part of the Solution

The system depicted in Figure 14.1 is called *The RCM Solution* because it *solves* the dilemma of whether full-blown RCM or a less robust process should be used. The answer is that both can responsibly be employed. The RCM Solution includes True RCM while also incorporating two other processes that capitalize on the versatility of RCM principles: *Reverse Engineering Method (REM)* and *Lubrication, Cleaning, and Servicing (LCS)*. Additionally, it incorporates a decision to run an asset to failure (RTF). The Analysis Selection Process (ASP) allows equipment custodians to determine which of these solutions is technically appropriate for each asset or system.

The concept of determining what systems and components warrant full-blown RCM isn't new.

The Concept of Deciding What Assets Justify the Application of True RCM

As previously set forth, the use of streamlined RCM as a substitute for True RCM is strongly discouraged; as a result, The RCM Solution may seem like a contradiction. It is not. In fact, the concept of determining what systems and components warrant full-blown RCM isn't new. The original architects of the RCM process, Stanley Nowlan and Howard Heap, didn't intend for RCM principles to be applied to all aircraft systems and components. This concept was detailed in their ground-breaking 1978 book entitled *Reliability-Centered Maintenance;* Nowlan and Heap describe an aircraft as:

> ...a complex system that is made up of a vast number of parts and assemblies. All these items can be expected to fail at one time or another, but some of the failures have more serious consequences than others. Certain kinds of failures have a direct effect on operating safety, and others affect the operational capability of the equipment....There are a great many items, of course, whose failure has no significance at the equipment level....The first step in the development of a maintenance program is to reduce the problem of analysis to a manageable size by a quick, approximate, but conservative identification of a set of significant items—those items whose failure could affect operating safety or have major economic consequences.

This reasoning is also used in MSG-3, *Operator/Manufacture Scheduled Maintenenace Development,* which is used to formulate proactive maintenance policies for prior-to-service commercial aircraft. MSG-3 states:

> Before the actual MSG-3 logic can be applied to an item, the aircraft's significant systems and components must be identified. Maintenance Significant Items (MSIs) are items fulfilling defined selection criteria for which MSI analyses are established at the highest manageable level. This process of identifying MSIs is a conservative process (using engineering judgment) based on the anticipated consequences of failure. (ATA MSG-3: Operator/Manufacturer Scheduled Maintenance Development, Revision 2009.1, Section 2-3-1).

The logic described by Nowlan and Heap and outlined in MSG-3 can be applied to any organization that has a multitude of assets and systems to manage such as a manufacturing plant. It too is a "complex system that is made up of a vast number of parts and assemblies" that vary in size, technical complexity, how each piece of equipment serves the plant, and how much that service matters. Some of the equipment may play a critical role in production whereas others only offer support roles. Just as an aircraft can be divided into systems so that it is possible to determine which systems warrant RCM, a manufacturing plant's equipment can be arranged using the same logic. Figure 14.2 is an example of how equipment associated with a plant air system can be partitioned.

By using this technique, assets are arranged in descending order of complexity.

Analysis Selection Process: Determining the Right Level of Rigor to Apply to Each Asset

The Analysis Selection Process (ASP) is applied at the asset level. As detailed in Figure 14.2, assets are partitioned to a manageable level so that equipment custodians can establish what assets to put through the ASP. The ASP determines what level of rigor to apply to each asset, as depicted in Figure 14.1. For example, the following assets are likely to be put through the ASP: compressor, air dryer, tower water distribution system, lathe, band saw, forklift, brake press, elevator, fire hydrants, sprinkler system, etc.

The ASP contains defined selection criteria in the form of questions that are arranged as an algorithm. The algorithm leads the user to one of four analysis tools: RCM, REM, LCS, or RTF (Run to Failure), based on the answers to the questions.

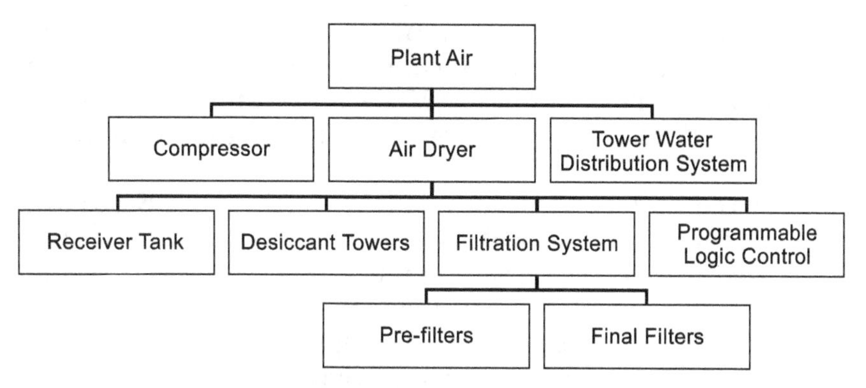

Figure 14.2 Partitioned Plant Air System

Examples of questions embodied in the ASP may include:

- Is the current level of safety acceptable?
- Is the current level of environment integrity acceptable?
- Is the current level of operational capability acceptable?
- Is the current level of system reliability acceptable?
- Does failure have an unacceptable economic impact?

It is important to note that the questions embodied in the ASP are formulated by an organization and are thus unique to that organization. There isn't a cookie-cutter solution because there are a range of variables from one organization to the next that influences the ASP. Some ASPs may be simple whereas others may be more complex. A trained facilitator can assist an organization in formulating an ASP.

Reverse Engineering Method (REM)

REM does not involve the level of rigor associated with an RCM analysis. It is important to note that, although it is based on RCM principles, REM is not RCM. During an REM analysis, the manufacturer recommendations—or a suitable template such as a legacy maintenance program—are reverse-engineered to identify what Failure Modes those tasks are intended to manage. Examples of this are depicted in Figure 6.7 of Chapter 6.

After the Failure Modes are reverse-engineered, Steps four through seven of the RCM process (Failure Effects, Failure Consequences, Proactive Maintenance tasks and associated intervals, and Default Strategies) are then carried out. In addition to this, all protective devices (such as relief valves and low pressure shutdown systems) are analyzed, applying all seven steps of the RCM process. It is important to note that REM only addresses protective devices and those Failure Modes that correspond to maintenance tasks that have already been established for the asset. Therefore, the solutions produced by REM are limited.

Lubrication, Cleaning, and Servicing (LCS)

LCS is a process applied to a physical asset if the rigor of RCM and REM are not warranted. The scope of an LCS analysis is limited to formulating lubrication, cleaning, and servicing tasks. Thus, the asset is analyzed from that perspective: only question six of the RCM process is applied. However, just like RCM and REM, all protective devices are analyzed applying all seven questions of the RCM process. Therefore, the solutions produced by LCS are very limited. Typically equipment such as lathes, drill presses, and metal punches are subject to LCS analysis.

Run to Failure (RTF)

An organization's ASP may reveal that it is technically appropriate not to perform any analysis on an asset. That is, the optimal failure management strategy is to do nothing and the asset should be allowed to run to failure.

Summary

It is important to note that the less robust processes—REM and LCS—are not called RCM nor is it suggested that they *are* RCM. They do, however, embody RCM principles.

Should full-blown, SAE JA1011 compliant RCM be applied to *all* assets? *Probably not.* Can less robust processes be applied to some assets and still be considered responsible custodianship? *Definitely yes.* However, organizations need to be systematic about determining what level of rigor to apply by ensuring that the consequences of failure at the asset level are understood and managed properly.

Appendices

Appendix A
Steps to Initiate an RCM Program

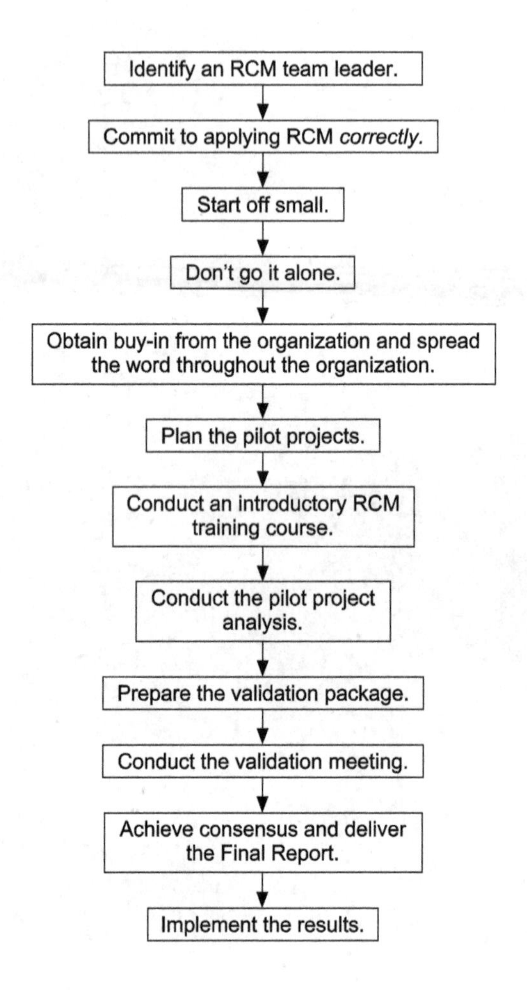

Appendix B
Steps to Successfully Complete RCM Pilot Projects

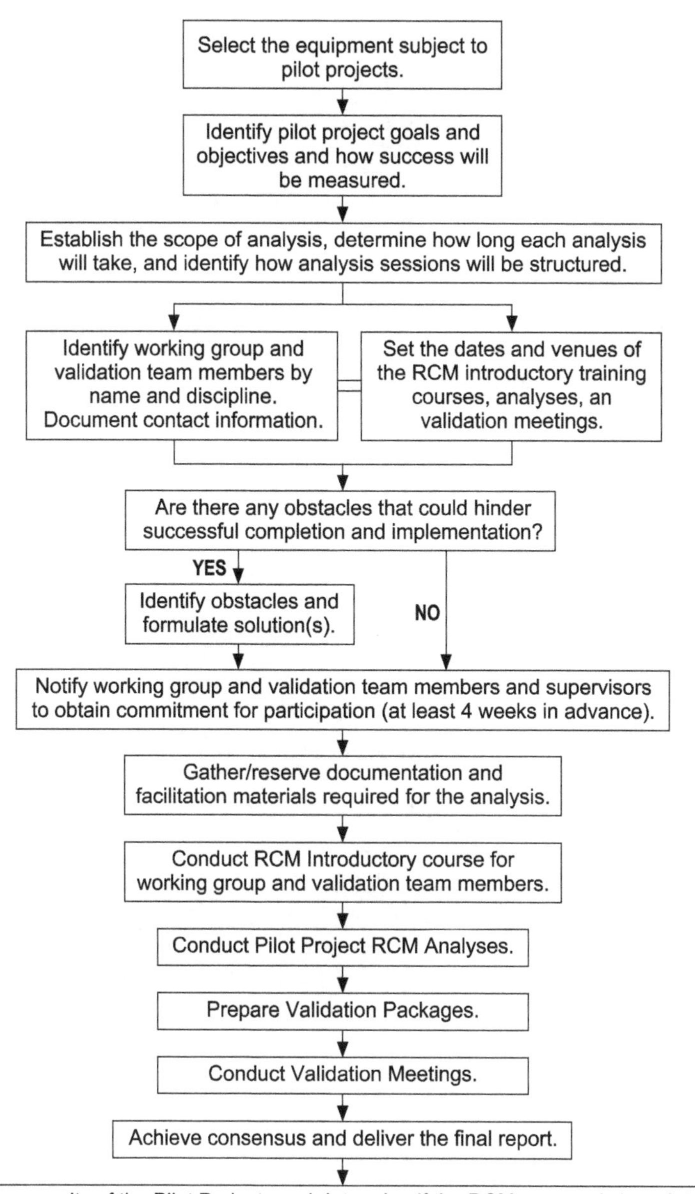

Glossary

Analysis Selection Process: An algorithm containing criteria unique to an organization used to determine which analysis process to apply to an asset: RCM, REM, LCS, or RTF.

Conditional Probability of Failure: The probability that a failure will occur at a specific age once the item has survived to that age.

Default Strategy: A failure management strategy, other than proactive maintenance, implemented to manage the consequences of failure (e.g. procedural check, Failure Finding task, no scheduled maintenance, engineering redesign).

Design Capability: What an asset, or any part thereof, is capable of doing.

Essence of RCM: To manage the consequences of failure.

Evident Function: A Function that, upon its loss, becomes evident to the operating crew under normal conditions.

Facilitator: An individual who is very well versed in RCM principles, leads the working group and validation team, and ensures that RCM principles are being applied correctly.

Failure Consequences: Describe how the loss of function caused by a Failure Mode matters; the categories of Failure Consequences are safety, environmental, operational, and non-operational.

Failure Effect: A brief description of what would happen if nothing were done to predict or prevent the Failure Mode.

Failure Finding Task: A task that checks if a protective device is in a failed state (e.g. blowing smoke at a smoke detector to ensure that it sounds an alarm).

Failure Management Strategy: An action taken to manage the consequences of a Failure Mode or a Multiple Failure (e.g. proactive maintenance, Failure Finding task, physical redesign, operating procedure).

Failure Mode: What specifically causes a Functional Failure.

Failure Modes and Effects Analysis (FMEA): A document that records the first four steps of the RCM process: Functions, Functional Failures, Failure Modes, and Failure Effects.

Failure Modes, Effects, and Criticality Analysis (FMECA): A document that records the first five steps of the RCM process: Functions, Functional Failures, Failure Modes, Failure Effects, and Failure Consequences.

Final Summary: A report that is compiled once consensus is reached between the working group and the validation team that details the final results of the RCM analysis.

Function: Describes the performance required of an asset.

Functional Failure: An unsatisfactory condition in which either some or all of a Function cannot be performed.

Hidden Function: A Function that, upon its loss, does not become evident to the operating crew under normal conditions.

Information Worksheet: A document in which Functions, Functional Failures, Failure Modes, and Failure Effects are recorded.

Lubrication, Cleaning, and Servicing (LCS): An analysis process limited to formulating lubrication, cleaning, servicing, and Failure Finding tasks that may be applied to a physical asset if the rigor of RCM and REM are not warranted.

Multiple Failure: Includes the failure of a protective device and another failure (e.g., low pressure switch fails and system pressure falls below normal levels).

Net P-F Interval: The minimum time remaining to take action in order to manage the consequences of failure once an On-Condition task is performed.

On-Condition Task: A proactive task performed at a defined interval to detect a potential failure condition so that maintenance can be performed before the failure occurs.

Operating Context: A document that includes a storybook identification of the system to be analyzed.

P-F Interval: The time from when a potential failure condition is detectable to the point that failure occurs.

Partial Failure: The inability to function at the level of performance specified as satisfactory.

Potential Failure Condition: Evidence that a Failure Mode is in the process of occurring (e.g., vibration, heat, cracks).

Preventive Maintenance: A scheduled Restoration or scheduled Replacement task that is performed at a defined interval without considering the item's condition at the time of the task.

Primary Function: The main purpose an item or system exists.

Proactive Maintenance: Includes scheduled Restoration, scheduled Replacement, and On-Condition tasks.

Procedural Check: A task performed at a specified interval to check for an evident failure that may have already occurred because it would be unacceptable to operate or continue operating with the failure condition.

Protective Device: A device or system intended to protect people, the asset, and the organization in the event that another failure occurs (e.g., smoke detector).

RCM Decision Diagram: An algorithm used to classify Failure Modes as evident or hidden, assess Failure Consequences, determine if any proactive maintenance is technically appropriate and worth doing, and determine if a Default Strategy is recommended.

Reliability Centered Maintenance (RCM): A zero-based, structured process used to identify the failure management strategies required to ensure an asset meets its mission requirements in its operational environment in the most safe and cost-effective manner.

Required performance: What an organization needs an asset, or part of an asset, to do.

Reverse Engineering Method (REM): An analysis process limited to analyzing protective devices and reverse-engineering manufacturer recommendations or a suitable template (such as a legacy maintenance program) to identify suitable failure management strategies.

Scheduled Replacement Task: A task performed at a specified interval that replaces an item without considering the item's condition at the time of the task.

Scheduled Restoration Task: A task performed at a specified interval that reworks or restores an item's failure resistance to an acceptable level without considering the item's condition at the time of the task.

Secondary Function: A function, other than the primary function, of a system.

Total Failure: Complete loss of Function.

True RCM: An RCM process that embodies the seven steps of RCM, which are carried out in order, as defined by SAE JA1011, *Evaluation Criteria for Reliability-Centered Maintenance (RCM) Processes.*

Useful Life: The age at which a significant increase in the conditional probability of failure occurs.

Validation Meeting: A meeting organized to review the working group's decisions to ensure they are safe and technically defensible and to determine what recommendations will be implemented.

Validation Package: A report prepared by the facilitator that contains all documentation necessary to perform a technical review of the working group's recommendations.

Working Group: A team of equipment experts assembled to carry out the RCM analysis; team members represent various disciplines such as a mechanic, operator, original equipment manufacturer (OEM), and a systems engineer.

Zero-Based RCM Process: The RCM process is carried out assuming that nothing is being done to predict or prevent Failure Modes.

Bibliography

The American Heritage Dictionary. Dell Pub.: New York, NY, 2001.

Hellenic Republic Ministry of Transport & Communications, Air Accident Investigation & Aviation Safety Board (AAIASB). *Helios Airways Flight HCY522 Boeing 737-31S at Grammatiko, Hellas on 14 August 2005.* Aircraft Accident Report, November 2006. Retrieved from: http://www.aaiasb.gr/Reports/AAIASB-R_2006-11en.pdf

MSG-3: Operator/Manufacturer Scheduled Maintenance Development, Revision 2009.1. Air Transport Association (ATA) of America, Washington, D.C.

Moubray, John. *Reliability Centered Maintenance.* Industrial Press: New York, New York, 1997.

National Transportation Safety Board (NTSB). *Safety Recommendation.* April 23, 2004. Retrieved from: http://www.ntsb.gov/recs/letters/2004/a04_29_33.pdf

National Transportation Safety Board (NTSB). *NTSB Advisory, Update on Investigations of Firefighting Airplane Crashes in Walker, California and Estes Park, Colorado.* September 24, 2002. Retrieved from: http://www.ntsb.gov/pressrel/2002/020924.htm

Bibliography

Nowlan, F. S. and H. Heap. *Reliability-Centered Maintenance.* National Technical Information Service, U.S. Department of Commerce: Springfield, Virginia, 1978.

Reason, J.T. *Human Error.* Cambridge University Press: Cambridge, 1990.

Regan, N. *Reliability Centered Maintenance Applied to the CH-47 Chinook Helicopter — Universal Principles that Go Far Beyond Equipment Maintenance.* Society of Maintenance and Reliability Professionals (SMRP) annual conference: Cleveland, OH, 2008.

Regan, N. *How to Initiate and Successfully Maintain an RCM Program.* SMRP annual conference: St. Louis, MO, 2009.

Regan, N. *The Power of Condition Based Maintenance Applied to DoD Systems.* Department of Defense (DoD) Maintenance Symposium: Phoenix, AZ, 2009.

Regan, N. *The Team Approach to the Development of Life Cycle Management Strategies.* ICOMS Asset Management Conference: Sydney, Australia, 2009.

Society of Automotive Engineers (SAE) International. Surface Vehicle/Aerospace Standard. SAE JA1011: *Evaluation Criteria for Reliability-Centered Maintenance (RCM) Processes.* Warrendale, PA, 1999.

U.S. Department of Defense. *NAVAIR 00-25-403: Guideline for the Naval Aviation Reliability Centered Maintenance Process (US Naval Air System Command).* Document Automation and Production Services (DAPS). Philadelphia, Pennsylvania, 2005.

Index

Index

Index

Index

Index

Index

Index